GU00692397

THE PIVOTAL MOMENT

The hurricane. The aftermath. The healing.

*The true story of a man who inspired
millions around the world*

KRISTLE BAUTISTA

THE PIVOTAL MOMENT © 2020 by Kristle Bautista.
All rights reserved.

Printed in Canada

Published by Author Academy Elite

PO Box 43, Powell, OH 43065

www.AuthorAcademyElite.com

All rights reserved. This book contains material protected under International and Federal Copyright Laws and Treaties. Any unauthorized reprint or use of this material is prohibited. No part of this book may be reproduced or transmitted in any form or by any means, electronic or mechanical, including photocopying, recording, or by any information storage and retrieval system, without express written permission from the author.

Identifiers:

LCCN: 2020908724

ISBN: 978-1-64746-272-7 (paperback)
ISBN: 978-1-64746-273-4 (hardback)
ISBN: 978-1-64746-274-1 (ebook)

Available in paperback, hardback, e-book, and audiobook

All Scripture quotations, unless otherwise indicated, are taken from the Holy Bible, New International Version®, NIV®. Copyright © 1973, 1978, 1984 by Biblica, Inc.™ Used by permission. All rights reserved worldwide.

Any Internet addresses (websites, blogs, etc.) and telephone numbers printed in this book are offered as a resource. They are not intended in any way to be or imply an endorsement by Author Academy Elite, nor does Author Academy Elite vouch for the content of these sites and numbers for the life of this book.

Some names and identifying details have been changed to protect the privacy of individuals.

Book design by JetLaunch. Cover design by Debbie O'Byrne.

Dedication

To the next generation

Christian and Connor,

this one is for you.

CONTENTS

FOREWORD

I have been asked, "Why, and how did you start Capstone[1]?" by many. One of those people was Kristle. I feel this book will give you a good understanding of the thoughts and process that made this possible. I first came to New Orleans for what I thought would be a week-long mission trip to rebuild houses after Hurricane Katrina. That was in 2009, and I'm still here today.

Kristle will tell you about the Facebook video that had over sixteen million views that generated over 900 emails in one night. To date, there are over 32,000 likes on the Capstone Facebook page. I never imagined Capstone would grow beyond one lot, and then the Capstone Ministry grew to operating fifteen urban farms on abandoned lots in the Lower 9th Ward and another two acres in Plaquemines Parish, Louisiana. Our first harvest of honey that year produced only three cups. Last year, we harvested over 250 gallons from sixty hives.

Since 2009, I've designed and helped to set up aquaculture systems for the Navajo Reservation and at a Ministry Center in Haiti. I've been requested to speak at multiple universities about food, faith and food justice issues. We've grown and given away over 18,000 pounds of food to those within our immediate community.

Kristle came to visit me again this year, three years after our first interview. She asked me, "What has changed since then?" My answer may sound like everything that has been accomplished is crumbling; I'll admit there was a time when I felt that way myself.

In September 2017, after a speaking engagement in Washington D.C., I had a bad fall that pulled the rotator cuff and biceps muscle completely off the bone. Near the end of physical therapy for my shoulder, I received the medical diagnosis with the discovery of two active types of cancer. In the past year, I had an abdominal mass removed for Follicular Lymphoma, and following that, I had my prostate and eighteen lymph nodes removed for prostate cancer. My Prostate Cancer Screening levels (PSA) continued to increase after the surgery, indicating that the cancer is still there. I'm currently undergoing radiation therapy today.

With my current physical limitations, we have downsized and returned some lots in the Lower 9th back to their original owners. I now have a handful of reliable volunteers who lead other volunteers, and I have a faithful assistant who helps me personally with Capstone daily. I told Kristle he may even be my successor. This was always one of the goals of Capstone—to empower people to provide for themselves.

The jail staff have ensured funding for those holiday meals to continue even if I'm not able to provide it. I still receive letters from some of the teens on occasion. I unfortunately haven't been able to join the youth in jail very many Sundays these days, but I look forward to when I'll have the energy to meet with them again.

That's quite a change from ten years ago when people would ask me, "Does it feel like home yet?" While I was trying to decide what to do at that time, a young associate had said to me, "David, stay where you're at so they can find you … and stay there until you are called to go somewhere else by God." I don't feel this is the ending. Maybe this is the final chapter or the beginning of another story, but then like so many things in life, I won't know until it happens.

Love, Peace & Unity,
David
New Orleans, Louisiana

PREFACE

Your life is made up of pivotal moments from the day you were born. And in these moments, you learn about who you are, who you are not, and are offered clues about who you are called to be. This calling exists within all of us, waiting for the moment you say "yes."

Welcome to David Young's incredible true story—how one man who had it all: the boat, the homes, the prestige and power—sold it all to help save the next generation. When you hear about people amassing more material things and wealth, know there is a man in Louisiana who believes your worth isn't dictated by what you have or where you come from. In the process of uncovering David's legacy, I unknowingly became transformed along the way too.

When the novel coronavirus slowly began spreading worldwide, David had given me the worrying news that his health had taken a turn for the worst. I flew to New Orleans right away, a few days after the Mardi Gras parade was over, and met with David in the familiar setting of the Lower Ninth Ward. Nothing could have prepared anyone for the global pandemic which was to take over the city of New Orleans in a lockdown against COVID-19 just a few short weeks later. The celebrations from Mardi Gras was believed to have fueled the massive outbreak in Louisiana. Suddenly all international travel came to a screeching halt, and daily news updates flooded our screens of the COVID-19 domino-effect never before seen in this lifetime.

The massive Ernest N. Morial Convention Center had prepared more than 1,100 beds as a field hospital with coronavirus cases spiking in numbers. Ironically enough,

the last time this same convention hall was used in an emergency was after Hurricane Katrina. Today, the world is being called to the same action that brought David to New Orleans in the first place, and his message of resilience in crisis more critical now than ever. Some have described these events as a time for reflection, a time for solidarity, for compassion, to help and pray for others on a global scale. This bittersweet reunion with my friend reminded me yet again that we all have a purpose, and when you talk to those who have seen the devastation first-hand and turned it into something beautiful afterwards, you begin to see possibilities rather than barriers.

We are living in unprecedented times, but wherever we find ourselves in the world, we are all in this together. I have always believed in the beauty of our dreams, especially in that pivotal moment where we are all called to the action that leads us there. We *can* make global change possible, one person at a time.

Kristle Bautista
Toronto, Canada

INTRODUCTION

A job description like David Young's doesn't exist anywhere, except within the heart. Perhaps it is part of a job description that we all share deep inside, without knowing that it's calling each one of us to apply. When it finally comes to the surface, it's that moment you realize you are meant to do something more—that pivotal moment when we're called to do something bigger than ourselves.

With visitors flying in from the US and all over the world (Russia, India, China, Japan, Australia, Thailand, U.K., Germany, the Netherlands, Denmark, Poland, Sierra Leone, Tanzania, and El Salvador), to see him, there were no signs of slowing down, as requests for his time to meet and speak were reaching in the thousands. The world needed to hear his story. This was my interview with David, and I wrote this book so you can meet him too.

CHAPTER 1

ON FATE

You have to keep breaking your heart until it opens.

—Rumi

This is it, I told myself in the mirror, *this is my breaking point.* Deep down inside, I knew I was going the wrong way. *There had to be another way.*

That morning, an email notification came up on my work screen, followed by a frantic call from an employee at one of our remote locations. "Hi Kristle," he started saying, "you don't know me, but Robin passed away suddenly last night, and we are all in a state of shock." She was a young receptionist who unexpectedly passed away from cardiac arrest leaving a young family behind and a team who simply adored her. I offered my sincere condolences assuring him that we would be there to support even from afar, and that I would be sending a grief counsellor to them the next day. It was a simple but supportive gesture to help ease the shock of a life gone too soon. As a new mother myself, the thought of her husband having to explain to their children why she wasn't coming home broke my heart.

The employee breathed a sigh of relief. "Thank you, Kristle. I really appreciate it."

My boss at the time was notorious for having a heartless approach, as evidenced by the revolving door of staff she was never able to keep long enough to appreciate. When she caught wind of my plan to send help, she demanded to see me to cancel the request, even though it cost the company nothing.

"If we send a grief counsellor to them," she said to me bluntly, "then we will have to send a grief counsellor every time someone dies."

The harshness of that moment shocked me as much as the complete lack of empathy. *Every time someone dies?*

That was the final straw among others, and I knew it. I submitted my resignation letter the following week. In those moments of turmoil at work, I realized that you have two options: you can either leave and find something else or stay and become toxic. There's certainly no in-between. And there are no toxic work environments, only toxic *people.* Most thought I was haste in resigning without a job lined up, but I couldn't let myself become desensitized to the humanity of others.

My boss was not a leader; she was a boss. What I learned that day was simple: If you find yourself in a dichotomy between keeping your job and being kind, then it's not a job; it's a prison.

Only a leader would know that thinking for yourself meant considering the world through another's perspective. Only a leader would encourage you to grow because your growth meant more for the company. And only a leader would want you to flourish because a boss would see your light as diminishing their own.

You'll know you found a leader when they want to see you become better and stronger than the day they met you. Like Steve Jobs famously said, "Your work is going to fill a large part of your life, and the only way to be truly satisfied is to do what you believe is great work. And the only way to do great work is to love what you do. If you haven't found it yet, keep looking. Don't settle. As with all matters of the heart, you'll know when you find it."

The three-week resignation notice period gave me relief more than anything. When asked about what my next step was, all I could say was that I was pursuing my passion in education. (It was no secret that I dreamt of the day I'd become a professor.) The departure gatherings filled my spirit with hope as many congratulated me on doing the right thing. Surprisingly, though, the one comment I heard the most was "I wish I could do what you're doing."

The only thing I knew with certainty was that there was another way. Experiencing roadblock after roadblock, disappointment after disappointment, I knew deep in my heart I couldn't stay there without a part of me dying every day. As in any relationship, if you can't trust the people you're with, it almost never works out. It was only a matter of time.

After my last day on the job, I immediately got sick with a nasty bout of the flu. It was likely the result of being in that toxic environment for so long, and needing to release the negativity that consumed everyone who joined that team.

Collapsed on the couch, wrapped in my robe and with a box of Kleenex nearby, there was no medicine needed or prescription required for this type of ache. I only needed time and rest to regroup.

It was an odd feeling, not having to rush anywhere, no meetings to go to or long days in the office. Now I was alone in a quiet house with only the hum of the refrigerator in the background, so I dusted off my laptop and opened a browser to Facebook. Scrolling through friends' updates of love and life, I noticed the exact same post shared by a few of my friends that morning[2]. It was a viral video of a Moses-looking biker—complete with a long beard, tattoo sleeves, and earrings—who was changing lives in New Orleans with his urban farms. Over sixteen million people were touched by the selflessness of this man named David Young, his video going viral within days of being posted online. Others had shared it over 200,000 times.

David was transforming people who would travel from all over the world to see him. People wanted to meet the man behind the mission. They wanted to see how one man's sacrifice and vision shone a light in the darkest of places and how he treated the most vulnerable and forgotten in our society. At that moment, I was captivated. How did he find such meaning in the middle of nowhere? How did he get there, and how did he find this purpose? There had to be more than what I saw in that video. The feeling of needing to find the light from a dark place spoke to me. In many ways, perhaps I needed saving too.

I suddenly felt this call to action. Having spent thousands of hours meeting with people of different walks of life at work, I knew that David's story was a special one. I felt empowered, uplifted, and inspired. I found his information at the end of the video and searched his organization online. I emailed him saying, "Hi, David. You don't know me, but my name is Kristle . . . I truly feel that you have a message that needs to be shared with the next generation."

Would he even respond? I wondered, but I shook off the hesitation and sent the email anyway. Within the hour, I saw a notification pop up on my screen with a direct response from him. My heart jumped when I saw it, and I opened it immediately. This was the beginning of a pivotal moment that changed my life.

Hello, Kristle. I'm glad you found my story inspirational. I have to say I am really surprised at the response it has generated . . . Your email stood out to me as one that I wanted to answer promptly because I feel you are sharing a good message that resonated with me. Attempting to remain humble whenever I get prompted, I have to share a few stories. Several years ago, someone who was volunteering and living with me, as many of the long-term volunteers have and still do said, "You're still raising kids; we're just in our 20s now."

He explained one aspect the video didn't capture was the teen jail ministry he had been leading since 2009. I realized there were even more people he was saving and kept reading intently.

When you say, "We live forever in those who learn from us," that is so true. I've had several people tell me I should write a book. My response has been I've already written the book. It's just scattered throughout many different people and venues. I simply need someone to collect and edit it. I am humbled when you say you would like to include me in your book. I look forward to talking with you further.

I couldn't believe what I was reading and felt tears coming down my face. What was sadness and frustration had now turned into tears of purpose and joy.

This might be the reason why everything at work was pointing me to leave, I thought to myself. And just like that, I booked my flight to New Orleans to interview him in person. In the video were short snippets caught on a cell phone, highlighting the urban farms David built on abandoned lots in the aftermath of Hurricane Katrina. I immediately began researching Hurricane Katrina, feeling guilty about not knowing more beyond what I remember seeing in the news.

Hurricane Katrina was an extremely destructive and deadly Category 5 hurricane that made landfall in Florida and Louisiana in August 2005. It caused catastrophic damage, particularly in the city of New Orleans[3]. To date, Katrina was one of the deadliest hurricanes ever to hit the US. During the hurricane and in the flooding that followed, an estimated 1,833 people died, and millions of others were left homeless in New Orleans and along the Gulf Coast. The damage was staggering. As of this writing, Katrina was the costliest Atlantic hurricane, clocking over $161 billion in damage[4]. The worst of this damage hit the Lower Ninth Ward in New Orleans, where David lived and worked. Katrina's victims tended to be low income and African American in disproportionate numbers, and many of those who lost their homes faced years of hardship[5].

After levees and floodwalls meant to protect New Orleans failed, most of the city was underwater, leaving most residents with homes they couldn't return to. All that was left behind were thousands of water-damaged structures, and the shell of a community once filled with life was left to the imagination of what used to be. In the loss, David saw the opportunity to create something, to give hope to the residents in the Lower Ninth Ward who had nowhere else to go. David's journey, I would later learn, began on the very first day he landed in New Orleans on a mission trip.

I traveled with my mom to meet him in person. Thankful to have my mom come along with me, I told her we were going on a girls' trip together. I told her about the Lower Ninth Ward and how devastated the area still was after the hurricane. Growing up in poverty herself, she was no stranger to hardship and can recall at any given moment the feeling of going to bed hungry every night with a tin roof over her head. Her heart broke so many times as a child that her heart bleeds for anyone who is in need of a helping hand or a warm meal. *If there's anyone who understands why I'm doing this*, I thought to myself, *it's her*. We landed in New Orleans at Louis Armstrong International, the first time for the both of us. I remember the look on her face once she saw the giant sign the limo driver was holding up as we were coming down the escalator. She started laughing and asked, 'Is this limo for us?"

We got into the shiny black Lincoln Continental and were on our way.

"Ever been here before?" our limo driver asked.

"No, we haven't. This is a first for both of us," I said.

He started giving us the tour of New Orleans, driving down the highway, past the historic cemeteries with interesting people buried there. He knew from the itinerary that we were staying in the French Quarter.

"So, what brings you here?" he asked.

"Oh, I'm visiting a friend in the Lower Ninth Ward," I said.

Puzzled, he looked at me in the rear-view mirror and repeated, "Did you say, the *Lower* Ninth Ward? Why would you go there?"

I started telling him about David and the viral post I'd seen the week before and how it inspired me. He paused for a diplomatic moment, and then said how utterly devastated the area still is.

He looked back at me again in the rear-view mirror, "And I haven't heard of this guy David."

"He's real. Trust me," I said. "You'll probably hear about him soon."

I knew we were pulling into the Lower Ninth Ward because everything completely changed. The blue FEMA hurricane storm tarps still covered roofs of a bunch of old houses, debris that must have been sitting there for years. I suddenly realized that we hadn't seen a single person around for blocks.

We slowly rolled onto his street of abandoned houses trying to find the right one. Most of them didn't even have house numbers. Our limo driver tried his best to guess which house it was.

I looked over at my mom. She said nothing, but the look on her face said it all.

"We're here," our driver said, coming to a stop.

I took a deep breath. "Let's go, Mom," I said. And in that moment, together, we stepped into one of the most dangerous neighbourhoods in New Orleans[6].

The house was propped up on grey concrete cinder blocks perfectly spaced apart. I heard the sound of chickens roaming freely all around, and I curiously looked underneath to see what else was there. It looked like a modular home that was delivered by a flatbed truck and then dropped off in the middle of nowhere. Most of the homes were a bright shaded colour—bright blue, bright red, or bright yellow. David's home was a bright orange. One of the chickens walked out from underneath toward the front where we stood. Recalling the rooftop helicopter rescues and the door-to-door emergency rescues on boats in the news, it felt like déjà vu, and here we were standing in the middle of it all. We slowly walked up the steps to the house with our bags in hand, knocked on the door, and waited. *Did I knock loud enough?* I wondered.

David opened the water-damaged and weathered wooden door, smiled, and extended a warm handshake to me and then my mom. I couldn't help but try to peek over his tall shoulder to see what was behind him. He looked exactly as he did in the video, like a biblical Moses complete with the long white beard. We walked inside the modest home that had many piles of paper and books in every corner waiting to be picked up, and he introduced us to the volunteers visiting that week. The house had fresh jars of honey ready to go in boxes at the front and an aquarium buzzing with different types of colourful fish. It had a welcoming feeling with people already busy working on their tasks for the day. There was a volunteer from the Navajo tribe, a volunteer from a local church group, and another student volunteer from Washington D.C. who was taking notes on the urban farms. They each stopped what they were working on to say hello and shake our hands, then promptly went back to what they were doing. Everyone looked like they were quietly on a mission.

First, we took a tour of the property. It was a surprisingly large property, about a quarter of an acre, factoring in the large urban farm and aquaponics and bee sanctuary he carefully crafted in the back. You would never imagine that all of this was hiding and operating in the backyard by looking at the small, humble house in the front. *What an interesting analogy,* I thought. *How many of us have so many unrealized dreams and hopes we carry in our hearts hiding somewhere in the back that no one would know but us?* He proudly showed us the urban farm, all the greenery, and what was fresh for the season. He mentioned that there were other lots he had taken over and rebuilt like this one, but we would only have time to tour this one property for now.

I was amazed at the rows of bees who were busy making honey, some of which he had rescued earlier that afternoon.

There were also goats, ducks, chickens, and fish. He said all of the goats had names, which made me wonder if he was joking. They're all named after celebrities, which I thought was hilarious. The large aquaponics pond he created in the back was impressive with different types of fish to provide nutrients for the plants. The waste from the fish is transferred through the water irrigation system and served as fertilizer for his crops.

"Have you built anything like this in the past?" I asked him.

"No," he said. "I YouTubed everything and figured it out along the way."

"Wow, that's incredible," I said as one of the goats came right up to me. I patted him on the head. "Do they all really have names?"

"Yes," he said, chuckling, "every single one of them."

Afterwards, the three of us went back into the house and sat down in the living room at an old wooden dining table with mismatched chairs that were noticeably missing the fabric that used to be there. The whole home felt like an open concept space, where we could smell the freshly caught fish that was being cleaned in the kitchen for distribution that day. It felt like there were tons of files and administrative requests neatly piled on the table. I learned these were from people around the world, all asking David for his time.

"OK, let's get started," I said, shifting in my seat, and asked him about those huge piles of paper organized by "requests to speak," "emails to respond to," and "visitation requests."

"Yeah," he said, nodding at them, "there's a lot. I'll get to that eventually." However, it was getting into hundreds of requests, thousands of emails. "I would consider hiring administrative help," he said, "but there is no budget for it at the time."

"So . . . are you from around here?" I asked.

"No, I'm from Indiana," he said proudly.

I glanced at his many tattoos, curious about his detailed sleeves of artwork. His arms were full of different images; it was like a giant mural on both sides. I decided to save those questions for later. I asked him how he manages such a large operation all by himself. He tells me about the variety in his day and that some days are more exciting than others, like rescuing honeybees early that afternoon right before we arrived. David worked on the farms six days a week from Monday to Saturday. Help would come in different forms and through volunteers from around the world who want to genuinely help.

"But what wasn't captured on that video," he said, "and what most don't know is that on Sundays, I spend it with teens in jail. I minister to them every Sunday on my day off."

Wait . . . he's a minister too? I thought.

Out of curiosity, I said, "Can I ask you something? You came here from Indiana to grow urban farms and still minister on your *one* day off?"

He nodded and said, "Yes."

"You just picked up and left everything behind?"

In my heart, I felt the words, *Ask him if he was a cop.*

So, I boldly asked, "Were you a cop?"

He looked straight at me for a moment, cracked a smile, and said, "Yes, I was. I was a chief of police in Indiana, serving the community for over twenty-two years before turning in my badge."

That was my *aha* moment. It suddenly all became clear. He understood the life of crime on the other side at the highest level. He saw the broken system and a broken world and decided to turn it around in the best way—perhaps the only way—he knew how. He started with his own life. He walked away from the only life he knew to give these kids a father-figure and a chance to make it. Even though the news wouldn't cover a story like his, he was a white police chief who was trying to help and make a positive difference in the black community.

We have seen the further impoverishment of Black families in inner cities, projects, and ghettoes as communities of color have become more desolate, more isolated and hopeless than they were in the '60s. We have seen the explosion of the prison industrial complex at rates that would've been unthinkable in the 1970s, with upwards of 2,000,000 men, women, and juveniles caged in American jails and prisons.

The United States, with only 5 percent of the world's population, has 25 percent of the world's prison population!

And for too many young Black men and women, the horror of incarceration has become a perverse rite of passage, marking one's transition from youth to adulthood. [7]

—Mumia Abu-Jamal, author of *Live from Death Row*

By the end of 2019, Louisiana was still considered the incarceration capital of the US[8] and long before David decided to live amongst some of the poorest in America back in 2009. For this reason, he would dedicate his life to bringing hope to these kids stuck in the jail system. He knew most of these kids had no real father-figure and needed someone in their lives to show them what a true mentor looked like. They needed someone who cared about them.

He had to make a life changing decision: to let everything go and literally start over in this unfamiliar city and build something out of nothing. He chose to give up all his possessions to do so. With only the modest funds he's raised from selling honey locally and small donations he would receive, he led his own silent movement that said, *This is our new generation, and these kids matter.*

Pivotal Moment in Action #1: What initiatives speak to you? What breaks your heart? The best causes are often the battles you face too. Is there a message you can send to thank someone or let them know how you feel? Is there something unresolved in your heart that needs your attention? Sometimes all it takes is that one message, that one moment where you say 'yes' to the calling, to begin transforming your life.

The things that resonate with you are speaking to you about something you're meant to do.

* * * * * * * * * * * * * * * * * *

It was Christmas morning: December 25, 1957. My mom was excited to finally open the one gift that was hanging in her stocking.

She reached into the old sock, only to find a small jar of Vicks VapoRub.

"Vicks?" she cried loudly. "This is my Christmas present?"

She remembers crying herself to sleep that night. There were no gifts under a tree, no warm Christmas dinner waiting for her at home. Their abject poverty in the Philippines struck her hardest on the one day she remembered hoping that Santa wouldn't forget her.

"My father completely abandoned us," she recalled bitterly, with tears welling in her eyes. At seven years old, she already felt as if she was carrying the weight of the world on her shoulders. Her older sister would be the one who shielded and protected her while their mother, a clothes washer, worked multiple jobs so they would have something to eat. Most nights they had nothing.

Her only memory of meeting her father was on the day he died. He waited just to see her and her sister before letting go. By this point, she was already in college.

"Did he say anything to you?" I asked.

She shrugged. "When he saw us, he started crying . . . and then we left."

"Did you cry?" I asked her.

"No," she said.

"You didn't cry at all?"

"Why would I cry?" she said abruptly.

"I didn't even know him."

* * * * * * * * * * * * * * * * * * * *

You can't go back and change the beginning, but you can start where you are and change the ending.

—C.S. Lewis

CHAPTER 2

THE CALLING

You don't write because you want to say something,
you write because you have something to say.

—*F. Scott Fitzgerald*

He never once spoke to me about religion but said he felt his actions would show you instead what he believed in. After learning that he needed to become a licensed minister to be put in front of the teens in jail, that's exactly what he did. For David, he knew actions always spoke louder.

He began to tell me stories of juveniles who were merely *days* away from being released, only to re-offend and be sent right back to jail. They would often receive even longer sentences than what they were initially brought in for.

"Why would they do this?" I asked him.

It was simply because these kids had no other options. There were no role models to show them a different life. Their fathers weren't active in their lives, and most of their mothers may not be in the right frame of mind—even if they were around. Many were drug addicts. The teens would describe being introduced into a life of crime because it was only their first juvenile offense, where for their older brothers or uncles, the same charge would potentially mean a lifetime adult prison sentence. As a family member, the trade-off seemed to make sense: *You take the hit for me, and I'll protect you.* With no other options, the street life made sense; there was respect, and there was family. For some, this was a blessing, and for others, it was an inescapable curse.

David knew these teens couldn't avoid their path, but he believed there was hope for a brighter future if he could show them a different outcome. His strategy was simple—to be a father-figure as best he could and show them love.

Every Sunday, he visited the group ranging from ages eleven to eighteen, sharing stories and talking about anything and everything. He made sure it was a safe place where he opened the door to whatever they wanted to talk about. He came with no agenda and imposed no expectations of what they should be talking about. He became like family to them, remembering their birthdays, sharing a meal on holidays. He kept track of their birthdays even when they moved on to different facilities.

On one of the visits, a teen asked him, "Mr. D, are those special glasses?"

"They're cheap glasses. Why?"

"I think they're special glasses, because every time you put them on to read to us and talk with us, you can see right through us."

For most of these kids, he was the only real friend they had. On the streets, you're expected not to talk to cops, but in his ministry, he was anything but a cop.

He tells me about another youth he has been visiting and talking to for nearly a year. This teen had only one month left to serve before being released when he told David that he "got bored and impatient" and decided to climb three twenty-feet high chain fences with razor wire at the top. He escaped, and they recaptured him within two hours. He spent forty-two days in solitary confinement on the adult side of the jail until he turned seventeen.

After that, at only seventeen years old, he was released into the rest of the adult prison population. All his juvenile charges were converted into adult charges, as well as adding new charges. As I heard this, I shook my head in disbelief. *He's just a kid.*

Once David located him in the prison system, he wrote to him and asked if he would tell him about the day he escaped jail. At that time, this teen was spending yet *another ninety* days in solitary confinement for fighting. He wrote back to David saying, "I'll tell you about it one day. Will you visit me?"

David responded with a letter saying that he would visit him and then asked if his fighting had anything to do with solitary being a safer place than being released with the general population. In his heart, David already knew the answer. In jail, this same teen would find ways to get into *just enough* trouble to get segregated from the others. He found jail a safer place than being on the streets.

After a few weeks, the visitation request was returned with the comment, "Inmate did not request visitation." *Bizarre*, David thought. Perhaps there was a delay in the paperwork, so he waited a month and sent it again. Once again, the rubber stamp declared, "Inmate did not request visitation." Puzzled, he sent him a separate letter and asked if he had even completed the request form? His reply was "Mr. David, I don't need to add you to my book (of approved visitors) for you to visit. Just knowing that you would is good enough for me. It's more than anyone else has done for me."

David was clear that while he didn't approve of the decisions these teens may have made, he knew that loving them had more to do with *how they need to be loved*, rather than placing conditions on them. It didn't matter what they did in the past. True unconditional love heals (rather than hurts) the other person.

He continued to tell me about the privatized prison system in the US, how Katrina brought him to New Orleans as a mission trip and then never looked back. The Lower Ninth Ward wasn't only in need of rebuilding homes, which is what he initially thought. It was so much more than that.

It's been 10 years since the watery carnage of Katrina . . . when the levees broke and the rushing waters of Hurricane Katrina swept into the neighborhoods of New Orleans, the Ninth Ward—the Blackest ward—received the greatest damage, and the least relief.

Today, 10 years after its horrific flooding, the Ninth is back to barely half of its former population of working-class and low-income inhabitants. It is but a shell of its pre-storm glory. What the residents of the Ninth learned was the hard, cold truth that they were all on their own, alone facing the fury of the storm. Oh, they could call 911, and they may even have gotten an answer, but no one came to help them . . . If we ever wondered whether Black Lives mattered, the squalid treatment of the people of New Orleans, especially the Ninth Ward, answered that question decisively.[9]

—Mumia Abu-Jamal

I began to understand why this was where David chose to be. These kids grow up in a world where they literally don't stand a chance. At that moment, I was all-in and wanted to help in whatever way I could. We started off as strangers, and in no time, we became like old friends catching up—except this story was unlike anything I've ever heard before.

Pivotal Moment in Action #2: Some people will love you, and others will hurt you, but all are meant to teach you about yourself. Who has entered your life and taught you meaningful lessons? Is there someone who has impacted you greatly, and wish you could tell them? Sometimes it isn't physically possible to talk in person —sometimes they have passed on or are no longer a part of our lives. However, these pivotal moments that stand out in your journey, both good and bad, have helped shape who you are today. Start by writing down the memories of those who have impacted your life, even if you never send it. One day, you might, but the important thing is to verbalize it for yourself and acknowledge the role they played in your life.

The teachers in our lives are everyone who show up and teach us something about ourselves. The teachers who have showed up for you, both painful and happy, have all contributed to making you stronger on your journey.

* * * * * * * * * * * * * * * * * * * *

I looked around me on the 7 a.m. United flight to Chicago. Most passengers wore suits or business-wear and were obviously heading to work. I realized I was the only one in a pair of Cons and a hoodie. I began buckling myself into my seat as the flight attendant was giving the routine safety protocol speech.

The plane began its take off. I looked out the window and thought to myself, "God, what am I supposed to say to David when I see him next? What are the questions to ask? Guide me please . . ."

The passenger sitting next to me sneezed. "Bless you," I automatically said and continued to look out the window.

"Thank you," he replied.

A few moments later he asked, "What brings you to Chicago?"

"New Orleans," I replied.

"Ah . . . you're not from New Orleans," he joked.

I laughed. "Right, I'm from Toronto. But how'd you know I'm not from there?"

"Because you said New 'Or-leens' . . . Anyone from the South would say "Naw'lins."

Good to know, I thought. If I'm going to keep returning, I should really learn some of the lingo.

I started to tell him that I was going to visit a friend who's into urban farming.

His eyes lit up. "Urban farming?" he repeated, as if I struck a familiar chord. "My daughter studied at Stanford and after graduating, took an apprenticeship under Will Allen (former

NBA player) from Milwaukee who started the movement called Growing Power."

In summary, Allen bought a two-acre plot after retirement outside of Milwaukee's largest public-housing project. Seeing that the main food sources in the area were based out of convenience and fast food, he created an urban farm, producing enough produce and fish to feed thousands. His daughter worked for Growing Power until she felt ready to start an urban farm of her own back in Wisconsin.

Incredible, I thought. Allen's story sounded so much like David's! I started telling him about David's journey, about capturing his story and how it was a calling that I felt compelled to respond to.

The man sitting next to me was Jeffrey Neubauer[9], the Democratic Party leader of Wisconsin from 1992 to 1996, who was also the founder and executive director of Higher Expectations with the vision of building a fully capable and employed workforce in Racine County, Wisconsin, and across the US. To accomplish that vision, they work from early childhood through post-secondary education with the goal of ensuring that every student—regardless of race, zip code, age, or family income—can succeed. His vision of helping the next generation made me see in that moment, I was right where I needed to be.

"You're a politician," I said. "May I ask your advice?"

I expressed my fear of not wanting to focus on the broken prison system in the US. The world didn't need more bad news.

"This is what you do," he smiled and in his fatherly wisdom said, "you tell David's story just as he wants it to be told, from his

point of view. Say it all—it won't offend anyone if it's coming from him. This is what people need to hear."

I suddenly knew what had to be done, and my purpose was clear. I thanked God in that moment for answering my prayer.

We landed in no time, and I promised I would stay in touch once the book was done.

Tell David's story just like he would want it to be told. The world needs to hear about more people like him.

And he was right.

* * * * * * * * * * * * * * * * * *

The best teachers show you where to look,
but don't tell you what to see.

—Alexandra Trenfor

CHAPTER 3

ON PURPOSE

There is a divine purpose behind everything—
and therefore a divine presence in everything.

—Neale Donald Walsch

There was a calming presence to David, a determined leader who never needed to exert authority to be respected. People loved him, people listened to him, and others simply wanted to be in his presence to learn. Inspired by his life-altering decision to leave it all behind, I asked him if he knew that this was his purpose. I wanted to know more about how he got there.

"How did you decide that this is what you wanted to do?" I asked.

"I responded to an ad that was posted in my local church in Indiana." he started. "It was a mission trip to help rebuild homes. I went for a few weeks at first, and then I came back a second time but this time for a month. Every time I came back, my stay would become longer and longer. I didn't even know how to build homes. I didn't know how to build urban farms or any of it. I just figured it out when I got there."

I said, "There's still so much devastation, and it's been over ten years? What happened to all the money? I remember seeing a bunch of charities and fundraisers, celebrities, Red Cross, FEMA . . ."

"All the money didn't get there," he said quickly. "And the money that *did* reach the community was tied in controversy and layers of procedures that many didn't have the ability to respond to. Residents were asked to provide proof of ownership for homes that they inherited over generations but never formally had paperwork. They simply couldn't prove that they 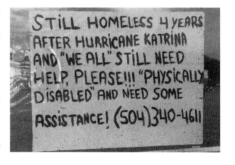 owned the properties. Politicians were being indicted for misappropriation of funds, reputable aid organizations raised significant amounts of money, only to be returned to the charity because of delays in the process. The money didn't make it to the Lower Ninth Ward where it was supposed to go.

"Mayor Nagin may still be serving time," David continued. "but the French Quarter and business district was rebuilt. Now, there's a high school and fire department here— it took *nine years* for that to happen. Only 30% re-populated back into New Orleans, everyone went around the country and the world, and some have since died."

By 2014, former New Orleans Mayor Ray Nagin was found guilty of federal corruption charges[10]. He joined a list of Louisiana elected officials found guilty of misdeeds while in office and was the first mayor to stand trial for public corruption. Nagin was the public face of the city during Hurricane Katrina, making national headlines as he blasted the federal government for its response to the storm and subsequent flood.

Of the twenty-one charges against him, he was convicted of twenty counts of bribery and fraud[11], some of which date back from before Katrina when he was in office for about three years.

I said, "So, the money that people donated thinking it was helping was just sent back? And what about the people?"

"People moved out, never to return," he said. "Some just picked up again wherever they evacuated to. Others came back to nothing."

Many don't have vehicles, so he drives residents to doctors' appointments to help them out. Access to healthy produce isn't easily accessible for most in the Lower Ninth Ward, where the closest grocery store is located miles away. While he has a mission to be there, he doesn't have a set daily agenda to follow. His task everyday is to serve the people in any way the community needs. And this varies from day to day.

Just then, a couple kids around ten years old knocked on the door, asking David for eggs.

"The eggs?" David asked, puzzled.

They responded, "Yeah, the ones that were under the house that the chicken just laid."

"Sure," he said, "you can have them. But I'm in a meeting right now." He promised to get back to them later in the day. He knew who they were and where to find them.

I joked that he's an egg farmer, and he said, "I suppose so too."

At this point, the August afternoon heat was getting intense. We decided it was a good time to take a break from the interview. David asked if I wanted to try feeding some of the fish. My mom, who was patiently sitting in the back, joined me in feeding the fish. He explained to us that nothing goes to waste; even the waste from the fish become the nutrients in the soil for the plants located right beside it.

Living in the city my whole life, I took for granted how easy it is to hop into the grocery store around the corner and have bountiful choices readily available. It was a beautiful thing to see what he created in many of these backyards, building something out of nothing.

I was in awe to be there, to see what the operation looked like in person. Before I left for the day, David handed me my first assignment. He brought out two journals, bursting full of letters people sent him from all over the world, testimonials, postcards, contact information written on little cards, promises to return, snippets of lives changed.

"I could try to tell you about impact," he said, "but you'd have a better understanding of what I do by reading others' experiences through their own words. I've never lent this out to anyone before. These are priceless to me, but I know I can trust you with it." I held on to those books and those inserts so closely on the way back to the hotel like you wouldn't believe.

Tired from the day's events of travelling and interviewing, my mom and I grabbed a quick gumbo dinner and called it an early night. I read through every single page/post and finally hit the bed at 2 a.m. while my mom slept in the next bed across the room. I wrote down the testimonies I wanted to ask him about, some of the interesting stories of their experiences. Many of them shared the same sentiment—that their lives were changed after meeting him.

Before I fell asleep that night, I remembered David telling me, "You felt called to come and see me based on a brief video that went viral. At face value, you are asking for one thing for the next generation, yet I feel personally you are seeking more."

He was right. I was looking for meaning in a world full of sad news. I was looking for inspiration and guidance and a purpose for why we are all here.

"God puts people in the places they should be and with the people they should be with," he said. The joy and purpose, he would discover, was in helping others.

To the outside world, it might appear like David took a step back in life after giving up his home and possessions. Taking the plunge and building urban farms would require him to let everything go. He was ready for the next level but not in a material sense. It also wasn't about rising in the ranks at work. We hear stories about heroes, and this time, I'd come face-to-face with one.

I was overwhelmed by the challenge of capturing his story, and even more so, that fate would bring this opportunity at the time I needed it most. I realized that we are all broken in our own way. None of our stories are perfect. But when we use our imperfections to relate to others and help, that's how we let the light back in.

Pivotal Moment in Action #3: Like it said in the movie, *Field of Dreams*[12], "If you build it, they will come." What is a problem that you can solve? What would you be willing to do to help? This could be offering a kind word to someone who looks like they are struggling, interest classes you can lead, being a mentor to the new person, or perhaps a letter you can write to a stranger. What comes easily to you might not be so easy to others and often are the gifts we don't realize we can share. Your gifts are meant to be shared with the world, no matter how small you think they might be.

The greatest accomplishments were often the simplest tasks, done with extraordinary care and love.

* * * * * * * * * * * * * * * * * * * *

I was walking by after an appointment when I noticed the sign "help wanted" in the window. This new car rental franchise must have just opened up I thought, as this wasn't there the last time I came home from school for the holidays. There was a tall man standing outside smoking a cigar when he saw me glance at the job posting sign. "Are you looking for a job?" he asked. He explained that he would need someone to help manage the day-to-day operations, work regular office hours, and would have the ability to drive any of the cars on the lot if they were available. He asked me how much I made previously at the airport, and I told him what it was. He said, "I'll pay you more than that." Without signing an employment contract, we shook hands and agreed that I'd start the following Monday. It was the perfect summer school job with convenient vehicle perks I thought, and couldn't wait to start.

We had a regular staff of five employees that I always ensured were paid on time. I managed the operations and signed off on the company cheques. However on Mondays, he would notify the team when someone new would be helping out temporarily for the week, mainly detailing the cars. The volunteers would work a nearly 8 hour shift like the rest of us, take lunches around the same time, except similar to a volunteer, none of them would be paid. Every volunteer that came in was diligent in reporting their time worked, ensuring that they did what was asked, never complaining. I finally asked one of them about why they chose to not get paid, and confided to me that they owed him money. He was kind enough to give them the option to pay the debts legitimately, and there were never any issues. Their respect for him was obvious, as it was with every visitor who came in who either knew him or simply came in to say hello.

My curiosity delved even further. Owed him money for what?

One day one of the volunteers asked me bluntly, "do you even know who he is?" Outside of this car rental, I admittedly knew nothing. That volunteer then said to me, "Kristle, let me explain it to you this way. If you ever got into trouble with anyone, I could help you. But if you got into trouble with him, no one can help you."

Over the course of that summer, I had more questions than answers. I wondered why he had to have a second accountant who would double-check what his other accountant was doing. Why there were people around all the time, always in a meeting. I had come to understand that the business was likely a front for something else, which I never did have the courage (or place) to ask. He tried to walk the straight and narrow, he would tell me, but couldn't work for anyone else. This was literally the only life he knew how to live. They took care of all their loved ones first. Their mothers didn't have to work. Their children and their mothers all had places to stay. Their many kids were decked out in the nicest of clothes.

But every single night, he wondered if that would be the day. I would ask, "The day when what?" I'd just get a look, and I knew what that meant. It meant that it was a fast life. They did everything they could for everyone else, knowing that it could very well be their last.

I was shocked to see the massive million-dollar drug bust all over the news years later.

Oh my God, I thought, this was my friend.

He was relentless yet forgiving at the same time, Jane and Finch which is the notoriously dangerous neighbourhood in Toronto he grew up in, groomed him for a life of hustling. In his mind, I realized he was an entrepreneur tasked with the burden of taking care of as many people as he could. This was truly the only life he knew how to live.

While I knew I'd likely never see any of them again, I became part of the family he trusted. He treated us all with kindness and respect. "Why not just keep a low profile, work a simple job 9 to 5, and then come home like everyone else?" I'd ask. He had a good heart, and we all knew that.

"Because that's not an option," he said. "Work a minimum wage job? And then what? This is the life I was given, Kristle," he said. "This life chose me."

"The wrong idea has taken root in this world and the idea is this: there just might be lives out there that matter less than other lives."

—Father Gregory Boyle,
Founder & Executive Director,
Homeboy Industries

CHAPTER 4

ON WHY

Any person or organization can explain what they do; some can explain how they are different or better; but very few can clearly articulate why.

—**Simon Sinek**

The next day, I brought back the journals David entrusted to me as promised. I placed the books back on the dining table and handed him a chai latte. I took a guess at Starbucks and realized we both love chai lattes by the smile on his face. The simple gestures brought him joy.

I told him how inspired I was by reading the journals and thanked him again for trusting me with them. He told me that he'd written letters to complete strangers and even wrote letters to people who have already passed on. His compassion for others always made him a great leader.

I mentioned one letter from a woman named Kathleen that read:

My life is the product of love and commitment.

I will forever know and have the love and support of the parents who gave me the gift of life.

Through the interaction of my life I have the power to influence goodness in others.

My life is blessed with the love and care of many people.

I am capable of sharing and giving love to others.

I control the paths my life will take by the values I choose.

I am a beautiful person with many talents that are specially mine.

I will survive the trials of my life with the faith and determination needed.

I have the right to demand fairness in all that challenges me.

I will hold my head high with pride for no one may take my dignity. Only I can surrender it.

Each challenge of life has been put before me, and nothing will be handed to me that I cannot deal and learn from.

I will make the effort to evaluate my values and never fear the knowledge of a new outlook.

I will love and respect myself first and foremost so that I can share a confident devotion to the duties and responsibilities set before me.

I will open my heart for the chance to grow and help others grow in spirit.

I have the ability to feel all emotions and develop compassion.

My experience is a treasure box waiting to open and give treasures to others in need of its gifts.

I may use the love of others in my search for inner peace.

I am worthy of being loved and demanding nothing less from those I give commitment to.

I am a special person who is loved, needed, wanted, desired, respected.

I am a responsible person who will challenge, learn, ask, open, give, and BELIEVE.

I am . . .

He said he didn't always know what people were looking for when they wrote to him. Sometimes like Kathleen, perhaps it was only for a friend to listen. "Whatever happened to Kathleen must have been tough," I said, "but it obviously made her stronger." She gained strength in her life through the trials, and by choosing to love herself, she is now able to help others.

Often strangers will send him stories about what happened to them, hardships they've experienced. I asked David about any hardships he experienced growing up. It was then that he revealed he was bullied as a child.

"Will you tell me more about that?" I asked curiously.

"I moved around a lot as a kid," he said, "and being the new kid, meant a lot of bullying. My dad was a district executive for the Boy Scouts. With his job, he ended up moving the family about every four years.

"At the time I didn't know that I was an introvert. I *did* know I had ADHD that was not treated, didn't make friends easily, was socially awkward, and was generally an average to below-average student."

As he reflected on his childhood, he shared an analogy of how things work in nature where the strongest survive. He told me that chickens have a pecking order, where every chicken introduced into the flock—even if it's a new chick from within—has to go through. All the chickens will peck at the others until it knows where it belongs within the flock. When it starts to peck back and find someone else it can peck on, it knows it has found its place.

"The bullying that took place," he said, "was mainly verbal with some physical elements. Most were either social exclusion, name calling, pranks, or being physically pushed around in the hall and class." He recalled a few times where the bullying led to physical altercations.

"I had just gotten a new three-speed bike—I know that will age me. It was my pride and joy. I rode it to summer school. Yes, I did summer school twice in order to graduate," he said. "Someone who I thought was a friend of mine grabbed it and said, 'It's *my* bike now' and took off riding it.

I ran after him, caught up, pushed him to the ground, and reclaimed my pride and joy. Later, I found out he had bloodied his lip when he hit the ground, and he was a bleeder. Now, in addition to the ongoing issues, I also had the guilt of hurting one of the people I thought may have been a friend."

"What some would consider insults and childish pranks when I was younger," he continued, "later became more intense as I grew older.

"I remember being the one they tied to a chair for two hours during knot-tying practice while the adult leaders chuckled on the other side of the room about how much fun they were having," David said.

"I remember having a pan of boiling water intentionally poured down my leg into my boot . . .

"Another time at school, I had been subjected to all the name calling and shoving I could tolerate," David explained. "One final insult with knocking my things out of my hands, and without thinking, I punched the person in the nose. I didn't know that I broke it, but it bled quite a bit.

"Once again, I had been pushed to the point of causing harm to someone else. The sad part is once I showed I could defend myself, I didn't receive as many insults or pushing around," he said. "Yet I also decided that I wasn't going to allow my frustration take control where I hurt someone again. I swallowed a lot of anger and only later in my life was able to find a positive and productive outlet for it."

He recalled feeling the discomfort of having to move as a child and it being a matter of "here we go again." While he didn't have good friends growing up, he did take comfort in knowing who *wasn't* a friend to him. "As a kid, it really is as simple as that. Either you are, or you aren't," he said. "Growing up is where things got complicated."

At this point, I started to gain a deeper understanding of how David became an advocate for those who have no one. He didn't have defenders growing up, he witnessed a lot of mob mentality, and was often picked on as the new guy or dubbed the weakling.

"I was smaller than all the others because I started school a year earlier due to all the moving to different schools. By grade four, I was diagnosed with ADHD. I was tested for behavioural issues as a kid, did all the brain scans, and was prescribed Ritalin. I went from failing classes to getting Bs. From sixth grade onwards, I stopped taking Ritalin, took it again for two years in college, and then again at forty-nine years old."

After he left the police force, he retested again, and his ADHD was off the charts in three out of four categories. When he was working, he was consuming so much caffeine that it triggered his ADHD to a point where the caffeine was no longer effective. Since then, he has learned how to focus and channel his energy, trying his best to limit caffeine and find ways to get off Ritalin. He took on focused-leadership roles and taught others how to treat people with respect. With his ministry, the teens in the jail are his inspiration, and the positive outcomes he sees in those who lack hope became his *why*.

David recalled the one teen who was in jail for six years, who became the unexpected leader that mentored other juveniles before he left jail. He came in as a broken child and left the jail with a message of hope for his fellow friends, telling them to have faith in themselves. It was in moments like this where David knew he was doing the right thing. He helped change the heart of one teen, giving him a real chance to survive.

"What happens after they leave?" I asked.

For some, he loses contact after they leave, some transfer to other facilities, some have probation where they're not allowed to be around others. Some pass away.

Many of them are trying to raise themselves even if they have a single parent or grandparents around. So, sometimes, they end up with the wrong influence around them that negatively impacts their decisions. It's an uphill battle. In trying to find supportive and positive people, sometimes they get rid of friends they've had their whole lives or even their family (if they're a negative or destructive force).

"So, what happens to them after they leave?" he repeated. "There's no general one-size-fits-all answer. Sometimes, it's trying to get them to think about what they need to do to ensure they don't return. Sadly, not all of them are even guilty but could spend years waiting on a decision to be made. Meanwhile, most of their best impressionable years as a child are spent wasted away. It's not always something new they've done. It could be something previous they've done with a backlog in the court system."

"This really is an uphill battle," I said. "Where do you even begin?"

"It all points to the same direction," he said. "They're all looking to be loved."

Pivotal Moment in Action #4: One profound piece of advice I received was to think about what others thank you most for. Really give this some thought. What do people thank you most for? This will guide you toward what you are strong in and what makes you unique. Understanding your natural strengths and why you do what you do naturally will help you break free of the negative opinions of others.

*The areas of impact that you are noted for are where your natural strengths lie,
regardless of what others think.*

* * * * * * * * * * * * * * * * * * *

"Mama?" my son Christian asked me while I was busy making Saturday morning banana pancakes.

"Yeah, sweetheart?"

"Who's the guy who helps people in jail?"

I paused for a moment. "David? He's the guy who helps people in jail."

"Why does he help people in jail?"

I paused and sat down at the table next to him. "He helps people in jail because they don't have families."

"They don't have families?" He thought about it for a moment.

"But you have a family, Mama."

"Yes, I do."

"Daddy has a family."

"Yes, he does. He has a family too."

"But why don't they have a family?"

"Sometimes," I said gently, "people don't have families. David helps them so that they know what it feels like to have a family."

He nodded.

"That's what we do, Christian. When someone doesn't have a family, we help them so they know what it feels like, too, right?"

He nodded again. "Right, Mama."

* * * * * * * * * * * * * * * * * * *

Kindness is loaning someone your strength instead of reminding them of their weakness.

—Anonymous

CHAPTER 5

ON LOVE

There is more hunger in the world for love than there is for bread.

—Mother Teresa

Do you remember where you were when your heart broke for the first time? Do you remember the pain of your heart breaking afterwards and every other time after that? Like most of us who carry years of hurt in our hearts, the muscle memory of our hearts also grows stronger over time. Maybe even more cautious and guarded in trying to not repeat the mistakes of the past. Some pain isn't something we bargained for, though, in what we feel just can't be fair.

What I've realized is that nothing will make you immune to the devastation of a broken heart. It's like some sort of rite of passage for every stage of life: how to love, how to let go, how to forgive, how to grow. Now that I have two young boys, I often wonder what I would say to my younger self . . . And the truth is, you only grow when you choose to rise from the heartache. The heartbreak will come. But when it does, what helps you free yourself from the pain to start healing?

Sometimes it's easier to sit in the pain than it is to work through the pain of letting it go.

I remember the pain of speaking my truth later in life, only to see the pain reflected back to me in my mother's eyes. I asked her why she was crying because none of what I told her had any hold on me any longer. And all she could say was "I wish I knew."

I'll always remember the pain of a love left unrequited because it was never meant to be mine. The love that burns you to your core because you realize your partner cheated on you. Or the feeling of the dreadful walk into the vet's office while holding my 22-year-old fur baby as she peacefully took her last breath so she wouldn't feel alone while she crossed over. Or the memory of sitting in a sales meeting while I could feel I was having a miscarriage. I did my best to finish the meeting because it was an important one, then cried to myself on the bathroom floor once I got home. Or when I got the call from the doctor's office telling me something was wrong and needing me to come in to discuss my options. In every single instance, I remember going through my days with a glazed look in my eyes, wondering where all the love in my heart had gone. You could read every book there is on loss yet still feel completely helpless when it comes knocking on your door.

"The darkest nights produce the brightest stars." – John Green

Too often, we carry these silent reminders of why we might not be worthy of a love we would be willing to give. Sometimes things are not meant to be, or life isn't fair, people would say. But shouldn't love be one of those things exempted? Shouldn't love be a safe place, where we don't get hurt, where we feel at peace? The hard truth is, the more we love, the more we open ourselves to being hurt by others. The deeper we love, the deeper the pain of rejection and loss. But if we don't allow ourselves to feel this, how will we ever truly know what's real and what isn't? It's like the saying, "The darkest nights produce the brightest stars.[13]" If we are never mistreated, how will we ever be called to choose forgiveness?

When I was in my twenties, I was invited to a singles retreat, which I can admit that I didn't want to attend. What could they possibly teach me about love that I didn't already know or feel? A close friend referred me, saying that it was her gift for me to attend, assuring me that it would be good for the soul. So, with nothing to lose and a free weekend to spare, I brought my hesitant and guarded self. *God, I hope they don't make me say anything*, I thought. I hated the idea of having to introduce myself to a sea of strangers, let alone talk about the hurts that had paralyzed me for years.

On the retreat, we were instructed not to bring any electronics whatsoever. No phones, no Internet, and there were no televisions, nada. We

"Only we can unpack what was put in our bag to carry"

were in the middle of what felt like nowhere, in these log cabins, with our duffle bags for the weekend. It felt like we were back in high school when "lights out" was announced, but everyone knew who had snuck in weed or alcohol to make it through the weekend. We related to each other in our similar disdain of being somewhere unfamiliar and feeling mutually lost. As the weekend progressed, we learned more and more about one another and the pain we all carried around like a backpack that doesn't leave us. Only we can unpack what was put in our bag to carry.

The following day, the pastor who led the retreat asked us what we knew about love and the different levels. He told us about the difference of the love between people and the love God has for us. The love God has for us, he explained, is a giving love that is unconditional.

Then, someone raised their hand and asked, "What happens if I give another person a hug, but I have feelings for them? Is it still *unconditional* if I want something more from them?"

Wow, what a question, I thought.

I'm certain the pastor's answer inspired most of us present in the room. He thought for a moment and said, "There's a difference between a love that *takes* and a love that *gives*." He continued, "When you hug someone to give them love, you are giving to them. But when you are giving a hug to take something from them, you aren't giving them love. You're taking it for yourself. You will always know the difference."

He went on to explain the famous bible verse on love. He asked everyone, "Who knows this verse, 'Love is patient, love is kind . . .?"

Everyone either raised their hands or nodded in agreement. He said, "God's love for us is perfect because He is love. We are all created in His likeness, and therefore, if you ever feel like you aren't worthy, that is a lie. We are all worthy because you are His."

Then came our first exercise.

On the screen, he pulled up the Bible verse he mentioned . . .

⁴ Love is patient, love is kind. It does not envy, it does not boast, it is not proud. ⁵ It does not dishonor others, it is not self-seeking, it is not easily angered, it keeps no record of wrongs. ⁶ Love does not delight in evil but rejoices with the truth. ⁷ It always protects, always trusts, always hopes, always perseveres. (1 Corinthians 13:4–8, NIV)

In the exercise that followed, we replaced the word *Love* and rewrote the verses with *our own names*. I can't tell you how powerful that moment was. I began writing:

Kristle is patient . . .
Kristle is kind . . .
Kristle does not envy . . .
Kristle does not boast . . .
Kristle is not proud . . .
Kristle does not dishonor others, and is not self-seeking . . .
Kristle keeps no record of wrongs and does not delight in evil but rejoices in the truth . . .
Kristle always protects, always trusts, always hopes and always perseveres . . .

And the most important and the toughest one for me to write was the last one:

Kristle never fails.

At that moment, I felt this sort of redemptive path from wandering aimlessly to worthiness and peace. We are all worthy of love, no matter what has happened or what was done to us, and it all starts with you.

Pivotal Moment in Action #5

We are all worthy of love, no matter what we've done or what has happened in the past. Take a moment to empower yourself and align to the gift of loving yourself and others. Like that moment where that pastor asked me to write my name in the place of *love*, find a quiet moment and write yours down with pen and paper.

(Your name) is patient.
(Your name) is kind.
(Your name) does not envy.
(Your name) does not boast.
(Your name) is not proud.
(Your name) does not dishonor others and is not self-seeking.
(Your name) keeps no record of wrongs and does not delight in evil but rejoices in the truth.
(Your name) always protects, always trusts, always hopes and always perseveres.

And the most important one:
(Your name) never fails.

You are loved, and you are love. Now is your moment to shine.

* * * * * * * * * * * * * * * * * * *

"David?" I asked. "Those six full journals you wrote and threw away . . ."

"What about them?"

"Why did you throw them away rather than keep them?"

I guess what I really wanted to know in that moment was who he was writing about. It would've been the equivalent of him writing two novels.

He thought for a moment. "I just wrote it and threw it out," he said. "You have to love yourself before you can love someone else."

Sometimes, I suppose, when they say to love is to let go, perhaps moving forward means letting go too.

* * * * * * * * * * * * * * * * * * *

Our deepest fear is not that we are inadequate.
Our deepest fear is that we are powerful beyond measure.
It is our light, not our darkness, that most frightens us.
Your playing small does not serve the world.
There is nothing enlightened about shrinking
so that other people won't feel insecure around you.
We are all meant to shine as children do.
It's not just in some of us; it is in everyone.
And as we let our own lights shine,
we unconsciously give other people
permission to do the same.
As we are liberated from our own fear,
our presence automatically liberates others.

—Marianne Williamson

CHAPTER 6

ON LONELINESS

I used to think the worst thing in life was to end up all alone. It's not. The worst thing in life is to end up with people who make you feel all alone.

—*Robin Williams*

Nations around the world are facing a loneliness epidemic. Today, loneliness is deadlier than obesity and is only now being considered a major public health hazard. A 2017 study involving nearly four million people was conducted in the US, where researchers looked at 218 studies into the health effects of social isolation and loneliness. What they found was shocking. They discovered that lonely people had a 50 percent increased risk of early death, compared to those with good social connections[14].

If you are lonely, you are more likely to die an early death.

We are living in a time where we can reach out to people at an instant, but many are feeling more and more alone. So, what happens then to the teens and juveniles in solitary confinement? They are locked up for hours upon hours, with very limited social interaction, many of whom also have no loved ones to visit them. As human beings, we need to feel love and connection. So many of today's youth here in our own backyard are forgotten, thrown away by society.

But isn't this our new generation? I wondered.

Most of these teens that David works with grew up without a father figure, many without the love of a family surrounding them. They are labelled as juveniles and delinquents by their choices, but how can they thrive or give what they weren't shown? After the devastation from Katrina, many lives were shattered. In the loss, David found a new and deeper love to help, especially these teens. This path wouldn't have been possible if he didn't open his heart to responding to their need.

Only a person who knows what it means to have suffered can truly feel the depth and suffering of others. The loneliness David felt as a child and the feeling of being displaced struck a chord that these teens gravitated toward. His tough exterior with his tattoos and earrings made him someone they could relate to. It isn't everyday where your guidance counsellor likely has more tattoo sleeves and artwork than you do. I decided to finally ask him about it.

You're asking about the small tatt on my hand? It holds the meaning of overcoming suffering.

Like any other symbol or tattoo, it has other meanings that some may misinterpret. It is the third letter of the rune alphabet, "Thurisaz" or "Thorn."

From the suffering aspect, I think a thorn is more appropriate. Like any tatt, the personal sentiment it contains is only known to the person who displays it or who it is shared with.

To me, it reminds me of the suffering I went through to get where I am. I also know I am not the only one who has suffered.

The cross on my forearm was the first tattoo I ever got, and it represents a cross necklace I received from someone who was like a second mother to me. I still have the necklace but don't wear it anymore, so I have the tattoo as a reminder.

If I had to select a textbook interpretation, it would be this one:

> You have the power within you to face anything. Fear nothing, for you have the authority to claim your destiny.
> Let no one deter you from your search for the truth. Spiritual authority brings power and it is up to you to use that power in an unselfish and loving way. Power can corrupt if you do not have a true and honest heart.

The heart I permanently wear on my sleeve is red and black. Instead of the heart being under a cross on an old wooden door, it's an old wooden cross over the heart. At the top of the heart, I have a dove.

When I first drew out the heart on my tattoo, I had this image of a big bright Valentine's Day heart and a somewhat generic cross. I knew what it meant to me and couldn't wait to see what it looked like. If you have had a tattoo that was custom designed, you know that at times, the tattoo artist will add their own aspect to it as they add shading and highlights.

When he got done, I had a bright red heart, but it also had some black shading around the edges. At first, I wasn't certain if that was what I really was thinking. However, the more I thought about it, the more I realized it was exactly what it was supposed to be.

A black door and a black heart. They both indicate something pretty cold and closed off from the world. That would be a pretty good description of me a few years ago but not now. Perhaps that's a description of the Lower Ninth Ward before Katrina. I don't know because I wasn't here, but that is certainly not the feeling I get now.

The red heart that shows from beneath the black, cold, dead heart is filled with the love of God and Christ. In both cases, it took a disaster to start peeling the black away, allowing the red to show through.

Everyone knows a dove is the symbol for peace. While my dove carries an olive branch, it also carries my heart. It's a reminder that we're supposed to share our love of God and Jesus and extend that peace to others. In other words, the love of God and Jesus is supposed to be an open door, available to be shared with everyone.

This would be the only time David spoke about the love of God and how he became transformed. His heart was closed off to the world, and it took a catastrophe to open his heart again. Symbolically, David also wanted to talk about the door he chose for his organization in the Lower Ninth Ward. The door was badly damaged from the hurricane, tossed away as trash. The door had elements of the dove and colours of the black and red heart, like his tattoo. The locks no longer worked, but they weren't needed because it simply became an open door. Through the worst of it, this door remained strong and resilient. It reminded him of the Lower Ninth Ward. It became a symbol of strength and love open to anyone and everyone.

"This door will represent something personal to everyone who sees it," he said. When I walked through that door on the first day I met David, that's when my own transformation began. And it started by taking that very first step through the doorway.

We are often challenged to step into the greatest version of ourselves after the most difficult times in our lives. This reminded me of the unforgettable and powerful Super Soul Sunday talk Oprah had with Shaka Senghor. Oprah described the interview with Shaka as one of the best conversations she has ever had[15].

In 2010, Senghor walked out of prison after serving nineteen years for second degree murder. One of the most profound moments he described was the moment he shockingly received forgiveness by the victim's family in the form of a letter sent to him. The victim's family said to him, "I forgive you, and I love you." It was a powerful testimony of love and forgiveness that changed his path forever, one that would soften even the hardest of hearts.

I forgive you, and I love you.

Shaka said he had never felt that kind of love, compassion, or empathy for himself or others before. He, like so many of these juveniles, was only a boy who made a poor decision under the circumstances. His incredible journey of rising above and using his pain and experience to guide and mentor others carried one important message: we are all worthy of love and redemption.

He understood what it meant to be broken, and that when we can see the broken child in every person we encounter, we allow ourselves to respond with love and healing. Receiving another letter in jail, this time from his son, he decided that he would become the father his son wished for and needed to have in his life, something that Shaka himself never had. Today, he is the director's fellow of the MIT Media Lab, a University of Michigan lecturer and author[16]. What once was impossible, became victorious and righteous with the power of love.

Pivotal Moment in Action #6: We hold back emotions. We regret. We think about death as merely a possibility and not a guarantee. Think about who you would write a eulogy or a letter to if you heard they suddenly passed away? Who would you rush on a plane to see if they were on their deathbed? Like David, he gives love to people who are lonely, abandoned, and forgotten. Who could you give love to if you made the time? What would you want them to know?

***Your one act of love and kindness to one person might be the one thing that saves them.
Be that reason, even if others disagree.***

* *

"*They wished that you wouldn't find love or happiness,*" *a medium told me directly.*

My eyes widened. "What, like some sort of curse they put on me?"

"Yes," he said, "and this was intentionally done."

"Why would anyone want do that to me?" I asked this renowned healer.

He was an old man who received a gift of clairvoyance and healing later in life. In his youth, he was a prominent and wealthy businessman before receiving a vision that came to him in the form of a dream. Prior to this dream, he said to me admittedly, that he was not religious at all. The message he received was that he had been given a gift of healing, and that he was to use this gift to help others. He awoke the following morning able to foresee future events, ailments, and diseases (of people and their family members, both near and far). Sensing my skepticism, it was through the grace of God he said, that he was able to intercede and heal others through prayer if that's what they wanted. He never accepted payment for any prayer or healing he offered to you, and outright refused anything that seemed like a donation. His gift was meant to be shared, never monetized.

After he received this prophetic vision, he left his family business, wealth, and worldly things. He then went on to live a humble and simple life with his wife selling his paintings at a local market in Toronto. They called him Brother Eli.

I wasn't sure how to repay him for his time. No one works for free, I thought.

"I do this to help others, not for payment. And it's not me who heals," he said, pointing upwards. "It's God who heals."

He told me details about my own family members both here and around the world, about specific people in my life, events that have happened, things no one could have ever known. It was as if he could see them right then and there. He told me that my mom needed to forgive her father. And then he told me about things to come. I hung onto every word he said.

"Why would they do this to me?" I asked him again. "I didn't do anything to them."

"Always pray to God," he said, "and He will protect you. There is nothing more powerful than God." Then, he asked me to do something that confused me more than anything. He asked me to pray for this person.

"Pray for them?" I pleaded. "Why would I pray for someone who would wish hell on me?" I honestly couldn't understand.

"Because what is coming for them is ten times worse than what they wished on you."

"Why would I stand in front of their own karma?" I said. "Didn't they do this to themselves?"

He said, "Because people like that need our prayers the most." He saw the darkness in their hearts.

"Pray for them," he insisted, because it was the only way to break the cycle.

Conflicted by his request, but empowered by his wisdom, I never forgot our conversation. It took me years to finally realize that he was right. People who wish to hurt others are hurting inside themselves. Only the light can drive out that kind of darkness.

The child who does not feel the love of the village will burn it down to feel its warmth.

—African Proverb

CHAPTER 7

ON HOPE

One of the greatest gifts you can give another human being who is going through adversity is hope.

—*Shaka Senghor*

"How do you give hope to someone who's facing years in prison?" I asked David. He said that hope was probably the most difficult subject he's had to talk about.

"Sometimes we are hoping and praying for perhaps the wrong things," he said. "I understand that when going through difficult times, you want it to be over quickly. And if it isn't, you get discouraged quickly."

He knew each of them wanted to be released from jail early, but that wasn't possible.

David talked about one teen who was released for a day to attend his grandfather's funeral. Prior to that, the jail had never allowed anyone to leave for the funeral of a family member. A rare exception was made because he was one of the more well-behaved prisoners, and they thought it was the right way to handle his situation. But does that answer his prayer of wanting to be released early? Maybe not, but it gave him a little hope because he got out for a day. This teen saw that as long as he was patient, he wasn't going to be truly locked up forever.

"What about situations where it's nearly impossible, inevitable for a sad outcome?" I asked him. "How do you give hope?"

Then, he gave me the example of hospice situations with families who have relatives suffering a painful and agonizing death.

He's often asked, "If there is a God, then why isn't this illness taken away?" They are praying so hard for health to be restored, but nothing is working. David's answer was that it might not be the *right* prayer. Perhaps we should pray that they are comfortable on their journey if they are ready to move on. What if we prayed for clarity instead to understand what might be our own selfish reasons that we want them to continue? Why not pray for what might be best for them and what we can do to help them in the best way we can?

What if it isn't about us, but rather about them?

It reminded me of the story I heard of a young man sitting in the barber's chair. As the barber was about to start cutting his hair, the young man mentioned the tragic events dominating the news earlier that morning.

The barber thought for a moment and replied, "You know, I don't believe in God."

Surprised, the young man replied, "Really? Why would you say that?"

"If there was a God, then why wouldn't He stop all these bad things from happening?" the barber countered.

The young man responded, "Well, that's just like say-ing, 'Why are there so many people with bad haircuts? Is it because there are no good barbers out there?'"

The barber said, "No, of course not! I can't force anyone to sit in my chair."

And the young man said, "Exactly. How can God inter-vene if we won't let Him in?"

For David, he could only reach these teens if they allowed him to. For some, it would take years to build their trust and nurture the hope for something greater than their circumstances.

"Were you searching intentionally for God?" I asked him.

"At the time, I wasn't looking for God; it took a lot of reflection afterwards. There was no instant recognition of, 'Oh yeah . . . that was God . . .' It was more of a 'What the heck have I just done?' he said.

He said it even took a lot of reflection to allow every-thing to unfold in the way that it did. He left his career as a police chief swiftly, but it wasn't until sometime later when he realized this was God's plan for him. This to him was a better path than what he was on before.

"What gives *you* hope every day?" I asked.

"Just being able to do what seemed to be the right thing for the people who needed it," he said. "I work on different lots, and building urban farms, sometimes the families who owned it come back and want to take it back, so I return it to them. Every day, I focus on the task in front of me and how I can help. I'm usually up at 5:30 a.m. every day.

"Typically," he continued, "I'll take care of the animals in the morning and then do any gardening tasks that need to be done. Sometimes it's preparing the ground, weeding, harvesting, sharing food. It's giving someone in the community a ride to the city for an appointment or picking up prescriptions, etc."

"It's a tremendous amount of work to keep up with, especially when people come back and want to reclaim their homes only after you've already fixed them up," I said.

He nodded in agreement. But it doesn't stop or discourage him; he just keeps going and focuses on the next property he can rebuild.

"How do you manage to stay afloat financially?" I finally asked curiously. *He can't possibly manage all of his finances by selling these small jars of honey*, I wondered.

He managed to survive with zero income for a couple years when he first started. He uses whatever he has from his personal funds but honestly lives day-to-day below the poverty line. When he filed his income tax for the year prior, he had made just over $6,000 in income for the whole year. He's received some grants and donations that he uses to produce and sell honey, and the funds also help the operation to share the food that they give away.

He was determined to stay clear of becoming a huge charity, with tons of overhead and different interests that might come into play. He saw firsthand what happened with the funds that didn't make it to the residents in New Orleans and refused to do the same thing. It truly is a complete gift of selfless love and sacrifice.

David shared that he still writes letters of hope to the teens who have moved forward to other facilities. When they have birthdays coming up, he'll try to bring attention to them, and he tries to write them a more personal letter of encouragement. Once their current life status changes (either by being released or transferred), he tells them he's still available if they need him and gives them his contact info. Some stay in touch; some don't. Some of them choose to write it to him in a letter format rather than talk about it in person. He'll include a message that will be supportive, and he'll offer a bit of help for their situation they're in—by sharing some words of wisdom or advice without preaching.

There was a lot he couldn't discuss with me based on non-disclosure and privacy for the teens, and he reminded me that juvenile info is not public knowledge. So, he didn't want to ruffle too many feathers.

"What do the teens ask for most?" I asked him.

"Interaction with someone who's nonjudgmental," David said undoubtedly. "They face so many stereotypes and don't need more of that. They just want someone to talk to." They do an open forum on Sunday afternoons where they do some Bible study and a bit of worship, spend a lot of time talking about their real life, what's on their mind and what they believe in.

"What breaks your heart?" I asked him.

"Seeing so much potential being lost, feeling helpless and abused, feeling that there's not ever a way out or a light at the end of the tunnel. Youth sometimes say that being in jail is better than being on the streets. Getting in trouble in jail and time in lockdown or solitary confinement can be safer than time spent in the general population. They just don't see any other options."

A lot of states have privatized much of the prison services, sub-contracted out to for-profit-business[17]. As long as the prison system is privatized, he doesn't see the system changing. It's a money-making machine. And unfortunately, the teens who get caught in the cycle don't have many options that help them get out.

It just so happened that the Sunday prior, the teens told him that they wanted to read something about hope in the Bible. They were looking for a deeper discussion that needed a more direct correlation other than simply "hope and trust in the Lord." The topic of hope came up from a fellow cellmate who disputed their belief in God, and the others didn't know how to answer it. They aren't even certain what hope is or how it applies to them.

Then, I told him about Hollywood star and entrepreneur, Danny Trejo, who was on death row, but the moment he made a promise to God was when his life changed.

> "I remember asking God, 'Let me die with dignity. Just let me say goodbye. And if you do, I will say your name every day, and I will do whatever I can for my fellow man."
> **–Danny Trejo**

Trejo, a self-admitted hellraiser in his youth, recalls starting a riot in the prison where guards were hurt. "We went to the hole and were facing the gas chamber," Trejo said. "And I remember asking God, 'Let me die with dignity. Just let me say goodbye. And if you do,

I will say your name every day, and I will do whatever I can for my fellow man.'[18]

Trejo did not get the gas chamber. Instead, he was released from prison the following year, on August 3, 1969. With God fulfilling His end of their agreement, it was time for Trejo to live up to his side, he said.

"I have been keeping that promise," Trejo said. Since that time, the actor has been giving back by living clean and sober and trying to instill values into youth today. He is a Hollywood star now but remains grounded in helping others and remembering where he came from. His hope in his future came from giving it all to God.

Giving it all to God is not for the weak. You will walk through the shadows holding onto your hope and faith and still question if you're doing the right thing. But if there is any way God can use your life, then say something. Even if you don't know what to say, you can start by just saying yes.

Pivotal Moment in Action #7: Based on the widely shared story by Billy Graham, a tribal elder was teaching his grandson about life[19].

> An old Cherokee grandfather is telling his grandson a story. "A fight is going on inside me," he said." It is a terrible fight between two wolves. One is evil—he is anger, envy, greed, arrogance, resentment, lies, and ego." He continued, "The other is good—he is joy, peace, love, hope, serenity, humility, kindness, empathy, generosity, truth, compassion, and faith. The wolves are fighting to the death.
>
> Wide-eyed, the boy asks his grandfather which wolf will win. The old Cherokee simply replied, "The one you feed."

Which narrative do you play in your mind? Despite the difficulties you face in life, which wolf do you feed?

*The wolf you feed is the one
who survives long after you are gone.
Let this be your legacy.*

* * * * * * * * * * * * * * * * * * *

I remember the first time I met a Holocaust survivor. It was an elementary school class trip in 1997 to witness a Holocaust survivor speak about the horror and memory of what happened to her in Birkenau Auschwitz. Remembering being ripped away from her family, she began speaking solemnly about the family she was never to see again.

A classroom full of typically loud and gregarious tweens was brought to complete silence the entire time she spoke. She told us about the coffee they were given in the morning, which was nothing more than dirty water. They knew what it was but didn't have a choice. She felt complete sorrow recalling every detail in vivid memory.

Then she slowly pulled back her sleeve to reveal the tattooed number that remained imprinted on her. Of all the sadness she described to us that day, this was the moment when she began to cry. The painful memory of being addressed as a number, the moment her humanity was stripped away. I felt like I held my breath the entire time she spoke, until she told us about liberation day. It would bring the hardest of souls to tears just to hear her again.

We were all invited to come up afterwards, and meet her up close. The reason she spoke to us that day was so that no one would forget what happened, to bear witness that it never happens again. So, when people say that hope isn't a strategy, I would disagree. Sometimes hope might be all you have to hold onto, and the strength of the human spirit which cannot be broken.

* * * * * * * * * * * * * * * * * * *

To teach is to touch a life forever.

—Anonymous

CHAPTER 8

MIRACLES

There are only two ways to live your life.
One is as though nothing is a miracle.
The other is as though everything is a miracle.

—*Albert Einstein*

David often didn't know where the funds would come from, but they would always arrive in time. The resources, people, and money, while not much, would be exactly what he needed.

"My garden started as a 4'x14' plot of land, and if people wanted to help, they had to bring their own seeds or plants that they wanted to put in the ground," he said. "Even as we have expanded, we operate with very limited funding. Helping others became easier for me once I realized I didn't need to change the whole world. I could make a difference in my part of it, then let the rest of the world see it."

He would say he never worries either; everything shows up when it's supposed to. Including me. As I looked out the window, I marvelled at the entire operation and asked him, "How would you describe a miracle?"

He said, "When I think of a miracle, the main thing I think of is how Jesus healed people. However, it doesn't stop at the physical healing of an ailment. These miracle healings were brought about by their faith that they could be healed by Jesus or God. Today, people feel more comfortable referring to a higher power. Sometimes, people were healed by the faith of others on their behalf.

"But I also see some things that people claim are miracles that take on too much sensationalism, at least for what I would consider to be legitimate," he said. "TV evangelists lining people up, saying a few words, smacking them on the forehead, them falling to the ground, and are miraculously healed every time? I don't buy that at all. Even Jesus had to try a second time with one deaf man. When I read the occasional article about someone who had cancer and without further medical intervention became cancer free? I would call that a miracle.

"When I meet with the youth in jail," he continued, "there are a few stories some have shared that resonate. One of the teens told me how he had been out with a group of his friends. They had done their thing and gone back to a house. The house was locked up tight, and everyone was passed out asleep.

"He woke up with police in the room and in handcuffs. The police started searching his room. He said he knew he was in serious trouble, because once they searched under his mattress, he knew they were going to find his gun.

"When they lifted up the mattress, the police officer asked, 'What is this?'

"He knew he was done. When they showed him what they found, it was just a plastic bag of coins.

'My gun was gone' he said, 'God must have moved my gun from under the mattress.' Whether you call it a miracle or not, you are not going to convince that young man that God did not intervene in his life."

Another teen who was sixteen at the time remembered when he got shot for the first time. He sustained a gunshot wound to the head when he was thirteen years old and survived. He remembered his parents being drugged or drunk enough that they didn't bother stopping him when he left the house with their loaded gun. He recalls being angry and leaving to confront someone. Even though he miraculously survived, he does not consider that the miracle. He said, "By the grace of God, my gun jammed and did not fire. Otherwise, I would have shot and killed that guy who shot me."

"Which is the greatest miracle?" David asked me. "That he survived being shot in the head; that his gun didn't fire and kill the other person; or that he acknowledged the grace of God as preventing him from killing another person?"

"Another teen felt that he was put here (in jail) by God to keep him from getting killed on the street. He even said that if he had not been picked up the night before, he would have been one of fourteen people shot in the French Quarter earlier that year. It is hard to deny that there is a much larger picture of purpose and God's hand in their lives," David said.

"You are a miracle for them too," I said.

"I'd say that my transformation from working full time as a police officer, to a full-time volunteer doing what I'm doing now may be one ongoing miracle with many individual God moments that took place along the way."

"Do you pray?" I asked him.

"I believe prayer is overstated. I love kids' prayers. They stand there and say, 'Hey God, can we have some sunshine today?' They're not lengthy—it's just talking to God."

That's how he explains prayer. You don't have to pray with a million people; it just has to be sincere. He feels there's a certain energy that's developed if you share what you're praying for with others, those who will support you in what you're praying for. When praying, simply go to your private place, and don't be like a clanging gong.

When he made the decision to come to Louisiana, he knew he didn't have the funds or the plan, but he silently gave the worry and the how to God. He suddenly received funds from his insurance company in the nick of time on a claim he had been waiting

"Even the smallest of gestures made with love, the little things, can make the world of a difference."

years for. It came as a result where he had agonized and prayed and reached a point when he asked God directly, "How are we going to do this?" He released any worry in the form of prayer, he opened himself up to the answer, and waited for the response. The insurance payment covered him for the next seven months. The answer he felt was to *"just go,"* but the form took several days to present itself.

"It is such a beautiful thing," I said, "when we release control of the outcome and believe that miracles do surround us every day. Even the smallest of gestures made with love, the little things, can make the world of a difference."

We know that every positive thing we choose to do, big or small, positively changes the outcome somewhere else. This was one of the greatest lessons I've ever learned: every single thing you do matters. That moment you hold the door for an extra second for someone else, that "bless you" after someone sneezes, the smile to the person who might need it more than you know.

Little miracles surround us, if we are willing to look and listen.

Pivotal Moment in Action #8: Everyday miracles surround us. The universe is talking to us through people, songs, events and everything around you. Are you listening?

***Ask for clarity in your purpose,
and the answers will appear***

God . . . if you can hear me, please say something to me, I said quietly in my heart as I began walking in the sea of downtown commuters.

Please, say something to me.

I remember my heart being heavy that day, even though you'd never see it on my face. A few minutes later, a man walking in the opposite direction stopped me mid-traffic, reached across with his arm to stop me, and boldly said in front of everyone, "Excuse me, do you know that God loves you?"

"Sorry, what?" I said, startled. This has never happened to me before. Or ever.

He repeated it again.

"Do you know that God loves you?"

His random comment stopped me in my tracks.

"Why would you say that to me?" I asked.

"Because I saw you and felt that you needed to hear that," he said.

We stepped out of the walking traffic and began talking. It was a moment I'll never forget.

"What are you, psychic?" I asked him stunned.

How could he possibly have known what I asked God just moments prior?

"I'm not psychic," he said. Believe it or not, though, he said he was previously a Buddhist monk. This man looked like he could have walked out of a GQ magazine, complete with a custom suit, tall frame, and refined composure. It was hard to imagine his transformation.

In subsequent conversations, he explained how he didn't know how to reconcile his material wealth with his need to do meaningful work to help others. In an extreme move, he left his oil and gas business to become a monk.

"A monk?" I asked, surprised.

He grew up in the mining industry with the family business in Western Canada. The search for something more and the capitalism associated bothered him, even when he was young.

His name is William Divine, CEO of the global oil and gas mining company, Discovery International, Inc[20].

"How did you find your balance?" I asked him.

Through prayer and meditation he realized that he could in fact work on global projects, as that would also enable him to fund non-profit initiatives in third-world countries while building essential oil and gas infrastructure. He could use his strengths in business to help others. That's when he found his true meaning and calling. He is now a film producer as well, aiming to portray the oil industry in a positive light, in which he travels to several Third World countries providing them low-cost energy solutions.

"Just like those little flowers blooming in the middle of the concrete sidewalk," he pointed out to me once. "Is that not a miracle?" he asked.

We are all meant to shine in our own way and in our own time. And that moment for me was a message and messenger I would forever be grateful for.

We were born to make manifest the glory of
God that is within us.
It's not just in some of us; it's in everyone.
And as we let our own light shine,
we unconsciously give other people
permission to do the same.
As we are liberated from our own fear,
our presence automatically liberates others.

—Nelson Mandela

CHAPTER 9

ON LOSS

Darkness cannot drive out darkness; only light can do that.

—*Martin Luther King, Jr.*

David knew he couldn't save everyone. Some teens would re-offend and often on purpose. Many felt doomed in their reality with very little option for a different outcome. Whether it was physical or emotional abuse, it was a cycle that continued to repeat itself. David never gave up, however, on showing them what a father figure looked like.

It reminded me of that powerful scene in *Good Will Hunting*. Will Hunting, (Matt Damon) the child genius who had a troubled childhood of abuse and neglect. He lived as an orphan, moving from one foster home to another. In the movie, you could see that his fear of rejection and years of trauma kept him from truly letting anyone into his life. His past also blocked him from seeing his true potential. Will Hunting kept getting thrown in jail for his involvement in fights and then would brilliantly represent himself in court. On the outside, he had friends, was strong and tough, but deep inside, he was just a scared and hurting kid. It wasn't until the moment when Sean (Robin Williams, his psychologist) pointed to his case file encompassing all of Will's years of abuse from childhood and told him, "This is not your fault, Will. None of this is your fault."[21]

Will, quick to respond, says, "yeah I know." Sean very softly says again, "This is not your fault, Will" and kept gently repeating, "it's not your fault." Will finally let himself release all of the years of pain and neglect he had bottled up, and his transformation began when he came face to face with the fact that while his past was horrific, it wasn't his fault. He needed someone else to remind him of his goodness, that he was more than what happened in his past. The tears in that moment felt very real for anyone who's been through trauma. That we are worthy of love and to feel love.

I realized David was like that teacher who encouraged his students, by allowing them to release their pain through discussion, through companionship and through kindness.

"How do you help the teens move forward from the past?" I asked David.

"I can tell them what they should do. But I would rather sit and discuss with them and ask them, 'What do *you* think?'"

I nodded. "Understanding that deep down inside, they know the answer. You're just helping them get there on their own."

"And sometimes offering support if you're not sure yourself, it can be challenging, and sometimes you can present other options for them to consider," David said.

"Have you seen hardened hearts? Can you share a story where this turned around?"

"I've seen several, some more profound than others," David said. "You can see it as it happens, their whole demeanor changes, what they talk about while still incarcerated, and how they view being incarcerated. There is a realization that they have restorative value. They realize that they are worthy of redemption. Even the staff at the teen jail acknowledge that the ministry is something they look forward to, something different than being in their cell. A lot of times, the ones who seem disinterested are also the ones who have something to say or be the ones to ask for a message to be repeated. They're taking lessons in, whether consciously or not."

"Does anyone visit them?" I asked.

"Parents or guardians are allowed once a week, but no girlfriend or siblings are allowed. It's very sad because not many have visitors at all."

"I can't imagine what that would feel like . . ." I said quietly. *That's why he's here.*

This reminded me of another story about the notorious Sing Sing prison in New York. It has been closed down for a long time, but the history of how it housed some of the worst inmates in the US during the 1930s still remains. Looking at the childhood of so many of these prisoners, most had very troubling upbringings. Many experienced horrific abuses at the hands of those closest to them. They were never taught what it felt like to be loved and lived a life hurting others. And then they met Katherine Lawes[22].

Perhaps you have never heard of Katherine Lawes. Katherine was the wife of Lewis Lawes, warden at Sing Sing Prison from 1920–1941.

Sing Sing had the reputation of destroying wardens. The average warden's tenure before Lewis Lawes was two years. "The easiest way to get out of Sing Sing," he once quipped, "is to go in as warden." In his twenty-one years, he instituted numerous reforms—and an important part of his success was due to his wife, Katherine.

Katherine took seriously the idea that the prisoners are human beings, worthy of attention and respect. She regularly visited inside the walls of Sing Sing. She encouraged the prisoners, ran errands for them, and spent time listening to them. Most importantly, she cared about them. And as a result, they cared deeply about her.

Then one night in October of 1937, news was telegraphed between the prison cells that Katherine was killed in an accident. The prisoners petitioned the warden to allow them to attend her funeral bier. He granted their strange request, and a few days later, the south gate of Sing Sing swung slowly open. Hundreds of men—felons, lifers, murderers, thieves— men convicted of almost every crime conceivable, marched slowly from the prison gate to the bier, reassembled at the house, and returned to their cells. There were so many that they proceeded unguarded. But not one tried to escape.

If he had, the others may have killed him on the spot. So devoted were they to Katherine Lawes, the woman who daily walked into hell to show the men a piece of heaven.

Katherine's strength was to see the men less as prisoners and more as individuals. Thomas Moore has said, "We can only treat badly those things or people whose souls we disregard." To treat people well is to honor their souls. To honor their souls is to understand what it means to love your neighbor.

In 1921, Lewis Lawes became the warden at Sing Sing Prison. No prison was tougher than Sing Sing during that time. But when Warden Lawes retired some twenty years later, that prison had become a humanitarian institution. Those who studied the system said credit for the change belonged to Lawes. But when he was asked about the transformation, here's what he said, "I owe it all to my wonderful wife, Katherine, who is buried outside the prison walls."

Katherine Lawes was a young mother with three small children when her husband became the warden. Everybody warned her from the beginning that she should never set foot inside the prison walls, but that didn't stop Katherine!

When the first prison basketball game was held, she walked into the gym with her three beautiful children, and she sat in the stands with the inmates.

Her attitude was: "My husband and I are going to take care of these men, and I believe they will take care of me! I don't have to worry!" She insisted on getting acquainted with them and their records.

She discovered one convicted murderer was blind, so she paid him a visit. Holding his hand in hers, she said, "Do you read Braille?" "What's Braille?" he asked. Then, she taught him how to read. Years later, he would weep in love for her.

Later, Katherine found a deaf-mute in prison. She went to school to learn how to use sign language. Many said that Katherine Lawes was the body of Jesus that came alive again in Sing Sing from 1921 to 1937.

Then, she was killed in a car accident. The next morning, Lewis Lawes didn't come to work, so the acting warden took his place. It seemed almost instantly that the prison knew something was wrong.

The following day, her body was resting in a casket in her home, three-quarters of a mile from the prison. As the acting warden took his early morning walk, he was shocked to see a large crowd of the toughest, hardest-looking criminals gathered like a herd of animals at the main gate. He came closer and noted tears of grief and sadness. He knew how much they loved Katherine.

> *He turned and faced the men. "All right, men, you can go. Just be sure and check in tonight!" Then, he opened the gate, and a parade of criminals walked, without a guard, the three-quarters of a mile to stand in line to pay their final respects to Katherine Lawes. And every one of them checked back in. Every one!*

This incredible woman who didn't know any of these men walked daily into hell to show them what true love and kindness looked like. I had tears flowing down my face while reading that story back in the hotel room. These kids similarly face a potential lifetime (not only in jail) of a lack of real love. David had to lose everything to help others find themselves. And then, I knew, wished even, that there were more people like David out there.

Real love changes even the hardest of hearts.

Pivotal Moment in Action #9: It is said that if a memory makes you feel a negative or visceral response, it is unhealed and may be holding you back from your true potential[23]. These are uncomfortable, unforgiven, or painful memories that replay constantly in your mind. No matter how terrible or painful it is, try writing it down. That was one of the hardest things I've ever had to do but in acknowledging the past, has helped me to release it. Let it hurt. Let it heal. *Then, let it go.*

Your truth might not be beautiful, but the beauty isn't in the pain—it's in rising above. Those who can sing after the rain don't forget the storm, but they know that the rainbow will eventually come.

I was trying to print out my boarding pass at the New Orleans International Louis Armstrong Airport when a woman who worked for Delta saw that I was shuffling between papers and my cell phone. I had a connecting flight with a different airline and needed a separate boarding pass. She stopped and asked if I needed help. We started talking about where I'm from and how I have a little guy at home that I needed to get home to. She told me to love my son and enjoy every moment with him. After a few moments, she repeated it again . . . "Enjoy every moment and really hug your son." I asked if she had any kids, when she began to tell me how she lost her son years ago. He was eighteen years old when he was gunned down in a drug deal gone wrong. From a mother to another, it was heartbreaking to hear that.

"I'm not mad at what happened. I can't change that," she said. "But I've forgiven it."

Saddened by what she just shared, I responded, "I can't even begin to imagine . . ."

This woman who lost her only son had ultimate forgiveness for the tragic way she lost him. She said that while he might not be here physically, she knew he was always with her. She could feel him around her and knowing that gave her a sense of peace. More importantly, knowing that he, too, is at peace is what comforts her.

I hugged her and said, "He's right here with you. Always."

We never know what someone has gone through and what unimaginable losses or traumas they carry. And yet somehow, those deep wounds we have are what allow us to truly see each other.

"The loss is immeasurable but so is the love left behind."

—Unknown

CHAPTER 10

ON TRUST

As I walked out the door toward the gate
that would lead to my freedom,
I knew if I didn't leave my bitterness and hatred
behind, I'd still be in prison.

—*Nelson Mandela*

David's unwavering trust in God allowed him to follow his path. His journey has helped countless numbers of people in finding meaning and purpose. He had no paid employees or a large budget to work with. But his impact was far reaching, and the community trusted him completely.

"As the chief of police, you literally just walked away?" I asked.

He nodded. "That was my job. That was what I loved to do. But just like that, I walked away. I took time—about a month—to soul search and for spiritual guidance. I initially came to New Orleans to help re-build houses, and then the urban farms evolved after that.

"But did I ever doubt my decision to move here?" he continued. "Yes. Did I always know I was doing the right thing? No. What was the hardest part? Much of it. After I left full-time work and became a full-time volunteer, I went to several different disaster rebuilding projects which included Haiti. I also went on a mission trip to El Salvador as a presidential election observer. In one year, I worked on nine different disaster sites. In a general sense, I knew that helping others rebuild following a disaster was the right thing for me. After about six months of volunteer work, I was asked to go with a group to New Orleans. My initial project time was one week. At the end of the week, the long-term leaders asked me to come back and stay longer. I went back to Indiana for a month doing other volunteer work and then returned to NOLA for a month. At the end of that month, the leaders asked me to come back and stay even longer for an ecumenical building blitz that would be about four months.

"During my free time, I found myself called to go to the Lower Ninth Ward where I spent time meeting and talking with people on their porches. I went back to Indiana and did some other volunteer work. While I was away from NOLA, I would get calls from people asking when I was going to be back or if it felt like home yet. I had reached a point where any place I was at felt like home, but I don't think I was ready to call NOLA home yet.

"By this time, I had sold or given away many of my personal possessions because of financial necessity, and then because it felt good to share with others and lighten the material burdens I had accumulated. Within a few weeks of my returning to NOLA for a longer term, the uncertainty and insecurity of it all came upon me. I was sitting in my house that was now almost completely empty with less than $100 in my hand. When you've had a successful career with good benefits, only having $100 can become frightening. Now, I'm pretty happy if I even have $10. Perhaps this can fall into one of those miraculous events. I knew I had to return to NOLA. I had nothing that physically bound me to go there other than I just knew I had to. I didn't have enough money for gas to get there let alone to live other than what was provided for at the volunteer house. If I did make it there to work for four months, would I ever be able to leave again?

"I sat in the front room and cried. I asked God, 'You've brought me a long way and through many challenges, but how do I do this?' After a moment, I had a calm go through me and an assuring voice said, "Just go. Just go."

"What was the hardest possession to let go of?" I asked.

He struggled with his house being in Indiana—because it was a symbol that he had a place to go to, even though he was travelling and going back in between. Eventually, it became a place he couldn't afford, and it was only a place he would use to change from one volunteer place to the next.

"I guess that made it official," he said. "There's something really symbolic about the last thing you let go being your house and jumping in with both feet. There's truly no turning back."

"What's the most valuable thing you learned on the job as a cop?"

"I learned a lot continuously but learned more in the last seven years once I got beyond the road enforcement, i.e., tickets and calls about domestic disturbances. Where you have an opportunity to get to know people better, that's when I became a detective, then sergeant, then chief. That's where I started to really become active in the community," he said.

As he got more involved in the community, he was asked by the governor's council in Indiana to reorganize a community organization. It was at this time when his biggest challenge and learning opportunities came, when he learned the difference between being someone who told people what to do (authority) versus someone who guided people to do what needed to be accomplished and motivated them to do it on their own (as a leader). He restructured community organizations such as the Governor's Commission for a Drug Free Indiana, Working Against Substance Abuse (WASA) (under the Governor's Commissioners for Drug Free Indiana). Another one was Against Alcohol Controlled Substances and Tobacco in our Neighbourhoods (AACTION). These organizations he helped reconstruct still stand today.

"Would you say you were happy during this time?" I asked.

"Up until I resigned? Yes, at the time, I was happy, had a good job, good wages. Altogether, I supervised about thirty people back in 2008. I thought I could do another five years as chief but then came the day that changed it all."

"Tell me about that day," I said.

"It was a day that only God could've put together. There was severe weather, the possibility of tornadoes, and all three weather systems were supposed to come down, all of which dissipated or went another way—none touched down in Indiana. Then, there were a series of calls, one after another;

it was just not a typical day. One of the least-prepared officers was working that day when the call to check on the welfare of an infant came in. I went into that call personally, and I knew exactly who that person was. After I visited the home, I decided that someone was going to jail."

"You saved an infant that day?" I said.

Describing the memory of this child deeply troubled David. He paused as he recollected the call and the difficult decision he had to make in removing this child from the home of someone he knew. As the investigation continued surrounding the welfare of this child, he kept hearing the same voice in his head asking, "Is it worth it?"

"What did you mean by, 'Is it worth it?'" I asked.

"It was more directed to other people, what we're doing to them," he said. "These questions all related to 'Is it worth it?' kept coming up for this call. The only thing I can say is that the answer is 'No.'"

This was his breaking point, I thought. He knew he couldn't do it anymore. In one sense, he was saving the life of an infant. In the same way, he knew that by removing this child, he was also destroying this family. *There must be another way*, he thought. A *better* way. He knew he had to show these kids who grow up in hopeless situations how to heal the wounds they carry. How to show these boys what it means to be a man. How to turn to God in spite of the hardship because He will show you a better way.

Just then, he took off the 22-lb gun belt, took off his badge, and took off his keys. He called one of his supervisors and said he's leaving.

"Just like that?" I asked.

"Yep, just like that," he said.

"Did you feel fear at that moment? That maybe you might regret it?" I asked.

"People often focus on what needs to be done without thinking about what their heart is calling them to do."

"Do you ever worry?" I asked.

"I try to limit worrying," he said. "When I have issues that I worry about, I'll think about the multiple outcomes and the paths to getting to those outcomes. It works best when I ask myself, 'What now?'"

I nodded. "Oprah famously said to focus on the 'next best step,' I said. "What does that look like for you?" I asked.

"Sometimes," he said, "it takes a little bit for people to put trust in God and ask, 'What now? What *is* the next step?'"

He never loses faith but sometimes becomes uncertain about how things are going to work out in the physical sense. But he's never worried; he trusts that everything will work out the way it's meant to.

"And what's next for these kids?" I asked.

"Well, when you're between age eleven to eighteen, you don't want to wait for anything," he said. "Trying to explain waiting is not a one-time or even one-topic discussion. Let me take it to the most recently related question I was asked, *I have been in and out of jail most of my life. Every time I come back, I pray to God that I get released. It doesn't happen. How can you tell me that God answers*

'It may be a man's reason, or it may be God's reason, but you are here for a reason.'

prayers?' Since 2009, every teenager I have seen here has said that same prayer every night and every day. I have yet to see the cells doors open like they have in Biblical times. If that prayer was answered in the form that they wanted, this jail would be empty. 'You're here for a reason,' I would tell them. 'It may be a man's reason, or it may be God's reason, but you are here for a reason.'"

"I guess it's helping them to see the bigger picture?" I asked.

"They just need someone to take the time to show them."

Pivotal Moment in Action #10: Are there people in your life who hold you back, who constantly fuel fear in your dreams or decisions, or tell you that you're not ready? Rewrite how the story ends because no one's voice should be louder in your head than your own. If there is something you believe in, trust in your process, and take the steps necessary to make it happen.

*Trust in your process. From this moment on, no one's opinion or voice should be louder in your head than your *belief* in what you are capable of achieving.*

. . . .

In high school, my Dad insisted on driving me to school every day. I would only later realize how important these 'teen talks' would be. Dad had a new daily topic to discuss, some sort of lesson he's learned or a hardship he wanted to tell me about. He was honest about the mistakes he made growing up, and wanted to make sure I was prepared for the inevitable obstacles that would come. The truth wasn't an option for him, he would say, it was the only way to live.

Different from my mom, my dad's side of the family in the Philippines were well known in their hometown. My grandfather was a prominent businessman who founded the first university in their area, with this school still standing proudly today. My grandmother and her sister were known for their love of classical music, his aunt was a renowned concert pianist who continued to play full concerto pieces well into her nineties. How my parents met was left up to fate, I imagined, in what was ultimately meant to be. My mom would be eternally grateful for the way my dad changed her life, but more importantly that he didn't see her poverty as a weakness. For richer or poorer, he admired her strengths in humility and leadership as a result of her hardship, and her unconditional love for others as she too was once forgotten herself. She was everything he needed that money or status couldn't buy.

But one of the most important lessons my dad ever taught me happened on a day that he didn't *drive me to school. Surprised that Dad wasn't downstairs in the kitchen waiting for me to leave the house, I thought to myself, That's odd? He left without saying bye? Interestingly though, he left an open newspaper laid out on* the kitchen table, neatly folded to a specific article. We haven't subscribed to a newspaper in years, I thought, so it quickly stood out. The first thing I noticed was the publication date, which was from weeks prior. The article was on the long-term effects of street drugs on the brain. One of my closest friends at the time was kicked out of their house after a hurtful argument, and turned to drugs to help make ends meet. Maybe Dad already knew.

The very next day, Dad drove me to school again, and no one spoke about the article. Eventually he started talking first. He said, "You know, your mom and I won't always be here."

I shifted uncomfortably in my seat and said, "Yeah, I know."

He continued, "Even though one day we won't be there, God will always be there. Always pray to God. Trust in the Holy Spirit. He will guide you."

All I could say was, "I know, Dad" smiled and then quickly got out of the car.

That day, he left an important message that would resonate with me for the rest of my life. Through a simple article he placed on a kitchen table, he told me that he couldn't choose my path for me. That no matter what I choose, or the direction I go in, the day will come when they will no longer be there. But when that day comes, he made sure I knew that I had the opportunity to do the right thing. And no matter what situation I found myself, there was always a way out, and that I was never alone. God would always be with me.

These talks would later become my inner voice of reason.

I left the car that day saying, "I know, Dad," but what I really meant was, "thank you Dad, for loving me enough to let me be me."

A hundred years from now it will not matter
what my bank account was,
the sort of house I lived in, or the kind of car I drove . . .
but the world may be different because I was important in
the life of a child.

—Kathy Davis

CHAPTER 11

ON INSPIRATION

The caged bird sings with fearful trill of the things
unknown but longed for still,
and his tune is heard on the distant hill for the
caged bird sings of freedom.

—*Maya Angelou*

When I first met David in the summer of 2016, the future
for those stuck in the prison cycle of incarceration seemed
like an impossible uphill battle. How can you really escape
your childhood upbringing coupled with a system that
profits off your time in jail? But then on that bright day
December 21, 2018, I read the news headline that The First
Step Act (initiated by then-President Barack Obama) was
passed into law by President Donald Trump. The First Step
Act was the most significant criminal justice reform legislation in years[24].

"This legislation reformed sentencing laws that have
wrongly and disproportionately harmed the African-
American community," Trump said. "The First Step Act
gives nonviolent offenders the chance to re-enter society as

productive, law-abiding citizens. Now, states across the country are following our lead. America is a nation that believes in redemption."

The First Step Act, which passed with overwhelming support from *both* Republicans and Democrats, is an important and positive step in changing the federal criminal justice system by easing prison sentences at the federal level. At this juncture, it didn't matter who was pro-red or pro-blue – both sides agreed that people deserve a second chance.

"it didn't matter who was pro-red or pro-blue – both sides agreed that people deserve a second chance."

The First Step Act allows thousands of people to earn an earlier release from prison and could cut many more prison sentences in the future; many of whom can return to their families to re-build and start over again. Reducing recidivism — which is the likelihood of a convicted criminal to commit crime again, would be achieved by providing formerly incarcerated individuals a second chance at life with opportunities and stable jobs upon release. Celebrities like Kim Kardashian West, who is now an active criminal justice reform supporter became inspired to action after this law was passed. Launching a rideshare app that will help people find jobs after being released from prison, Kardashian West's program will provide gift cards to former prisoners in order to help them commute to job interviews, work, and meet their family members.

The first woman to be released under the First Step Act in 2019 was Catherine Toney[25], who was sentenced to twenty years in federal prison in 2003. After serving fifteen years, through the First Step Act, she was granted immediate release in 2019. Like so many others, she has transformed her life while in prison.

"I am committed to helping others change their lives for the better. For the last several years, I've prided myself of being a role model to my fellow inmates," she said. "I encourage them to better themselves through education, spiritual, and vocational programming. I strive to lead by example by taking each prison task and assignment and performing it to the best of my natural God-given ability."[26] She has committed to rebuilding her life with her daughter and granddaughter with this second chance.

Even for advocates like Fulton Washington, who was sentenced to a mandatory life sentence for a crime he did not commit, received clemency from then-President Obama in 2016[27]. Heartbreakingly, he knew that his relationship with his children would suffer the most. During his twenty-one years in jail, he started drawing and painting to pass the time. Art helped him get through the tough years until he was finally granted clemency. He started by drawing on envelopes and postcards that other inmates would send to their families. His attorney asked him to draw from memory a sketch of a witness who would be able to corroborate his innocence, and the sketch he drew was so accurate that his legal team was able to find the person.

Through prayer, he realized that this was God's plan for him—that he should pursue art, teach it, and share it with inmates and their families. Washington taught others to paint and was also convinced that while he was innocent, it was his art that would get him out of prison.

Upon release, he has dedicated his life to being a present father and grandfather. While the criminal justice system has "impacted my children at all levels, and being torn apart from my family has had a lasting impact on their lives," he said, he also knew that there was a bigger purpose in his life. Today he is a shining light to so many still struggling in the system.

"I feel that God has blessed me, that he has delivered on his word, that through faith, patience, and hard work, you'll inherit what you deserve—I believe in that," Washington reflected. "A lot of times, I wanted to stop. I wanted to do what everyone else was doing [in prison]—like watching TV or getting into a bodybuilding program. But I had to help inmates and their families, to create a bridge with art, to try to hold those families together."

The Philadelphia rapper, Meek Mill, was released in July 2019 and has now become a symbol of criminal justice reform and an outspoken advocate based on the injustices he has experienced[28]. Based on a misdemeanor gun charge back in 2007, he was finally granted a new trial and a new judge in 2019, overturning his conviction[29]. The lone witness in his case, a Philadelphia police officer, was ultimately found by the department of lying and theft.

Mill's lawyer, Mr. Tacopina, said, "What happens now is he (Mill) begins his life. For the first time since he was a teenager, he's not under probation, he's not under any supervision. It's the first time in his adult life that he'll be able to go somewhere without asking permission."

Mill's story and journey has highlighted the challenges and injustices that millions of people on probation and parole face today. So many more are not in jail but still are not free.

Advocates like these have shown that it is possible to begin again, to come out of prison victorious, and to fight for those who are still lost in the system.

David would say that he can't change the world, but he can start from where he is. Helping these teens avoid the adult side of jail was his chosen Goliath to tackle. And it starts with mentors like him, who knew that what these kids needed most was someone who can show them a different path.

What inspires David the most?

It's that moment when he knows he has changed a life for the better—when one of the teens he ministers to is finally ready to start over. He knows the risk they face, the environment they must thrive in, and the uphill battle that awaits them upon release. But he holds onto the belief that one day they will have a chance to begin again.

"What happens when they've completed their time?" I asked David.

Upon graduation, he has a simple ceremony where he presents them with a poem about becoming a man. *How beautiful and symbolic*, I thought. When he first meets them, they're only kids. And upon graduating, he challenges them to step into the best person he *knows* they can be. To become a man.

The poem he gives them is called "If" by Rudyard Kipling[30]:

If you can keep your head when all about you
    Are losing theirs and blaming it on you,
If you can trust yourself when all men doubt you,
    But make allowance for their doubting too;
If you can wait and not be tired by waiting,
    Or being lied about, don't deal in lies,
Or being hated, don't give way to hating,
    And yet don't look too good, nor talk too wise:
If you can dream—and not make dreams your master;
    If you can think—and not make thoughts your aim;
If you can meet with Triumph and Disaster
    And treat those two impostors just the same;
If you can bear to hear the truth you've spoken
    Twisted by knaves to make a trap for fools,
Or watch the things you gave your life to, broken,
    And stoop and build 'em up with worn-out tools:
If you can make one heap of all your winnings
    And risk it on one turn of pitch-and-toss,
And lose, and start again at your beginnings
    And never breathe a word about your loss;
If you can force your heart and nerve and sinew
    To serve your turn long after they are gone,
And so hold on when there is nothing in you
    Except the Will which says to them: 'Hold on!'

If you can talk with crowds and keep your virtue,
　　Or walk with Kings—nor lose the common touch,
If neither foes nor loving friends can hurt you,
　　If all men count with you, but none too much;
If you can fill the unforgiving minute
　　With sixty seconds' worth of distance run,
Yours is the Earth and everything that's in it,
　　And—which is more—you'll be a Man, my son!

David chooses to fight a new Goliath every day, and to each day, he says yes.

He doesn't have children—but is a father to many.

He doesn't have a spouse—but is a companion to those who have no one.

He doesn't have material wealth—but gives the riches of the earth to all who need it most.

If you look closely, you might find him puttering about his garden. If you knock on the door, he might even answer it. Ask him to show you his garden, and you'll see the collage of vegetables and plants he tends to daily. Ask him to tell you a story about life, and you'll be amazed.

Except on Sundays, when the teens eagerly await their meeting with him, you know where he'll be.

. . . .

"*I know some of you are moving on and graduating,*" *I said to my students after the last class of the fall semester.* "*Some of you may be travelling, others going into the workforce—or soon to be.*"

Looking at the group of 125 students staring back at me, I couldn't have been prouder. I could see the potential, the desire to succeed, and the brightness of their futures. It was never really 'success' the students wanted to hear most about, but rather my own failures and learning experiences along the journey that got me there. It had been years since I resigned from the role which completely changed my direction and led me to David's story. Defeat doesn't have to define you, I realized, it's who you are determined to become in spite of it that does. And now I was as a university professor teaching about the impact of great management.

"*Every job you'll ever have will teach you something.*" *I said.* "*You will become stronger after every single one. Look for the lessons, because they're there. Especially the tough moments, because they will come and you will become stronger for it. And now, I can honestly say that one of the best roles I had ever been given throughout my whole career was one that I wasn't paid for.*"

"*Some things are worth fighting for,*" *I continued,* "*which will lead you to the next big thing you're meant to do.*" *The students nodded.*

And indeed, it did.

. . . .

It is when you give of yourself that you truly give.

-Kahlil Gibran

THE TEN STEP CALL TO ACTION

All of us are seeking the same thing. We share the desire to fulfill the highest, truest expression of ourselves as human beings.

—Oprah Winfrey

I set out on a journey to document David Young's transformation but in the process, became transformed along the way too. We can all help one another, even in the smallest of ways, even in the saddest and most impossible of situations. Hope exists. Redemption is real. And we are all worthy of it.

Bet on the person you can become. Believe in the grandest version of your life, no matter what the circumstances are. Like David, the right people will step forward if you believe and are willing to take action.

Pivotal Moment in Action #1: What initiatives speak to you? What breaks your heart? The best causes are often the battles you face too. Is there a message you can send to thank someone or let them know how you feel? Is there something unresolved in your heart that needs your attention? Sometimes all it takes is that one message, that one moment where you say 'yes' to the calling, to begin transforming your life.

The things that resonate with you are speaking to you about something you're meant to do.

Pivotal Moment in Action #2: Some people will love you, and others will hurt you, but all are meant to teach you about yourself. Who has entered your life and taught you meaningful lessons? Is there someone who has impacted you greatly, and wish you could tell them? Sometimes it isn't physically possible to talk in person —sometimes they have passed on or are no longer a part of our lives. However, these pivotal moments that stand out in your journey, both good and bad, have helped shape who you are today. Start by writing down the memories of those who have impacted your life, even if you never send it. One day, you might, but the important thing is to verbalize it for yourself and acknowledge the role they played in your life.

The teachers in our lives are everyone who show up and teach us something about ourselves. The teachers who have showed up for you, both painful and happy, have all contributed to making you stronger on your journey.

Pivotal Moment in Action #3: Like it said in the movie, *Field of Dreams*[31], "If you build it, they will come." What is a problem you can solve? What would you be willing to do to help? This could be offering a kind word to someone who looks like they are struggling, interest classes you can lead, being a mentor to the new person, or perhaps a letter you can write to a stranger. What comes easily to you might not be so easy to others and often are the gifts we don't realize we can share. Your gifts are meant to be shared with the world, no matter how small you think they might be.

The greatest accomplishments were often the simplest tasks, done with extraordinary care and love.

Pivotal Moment in Action #4: One profound piece of advice I received was to think about what others thank you most for. Really give this some thought. What do people thank you most for? This will guide you toward what you are strong in and what makes you unique. Understanding your natural strengths and why you do what you do naturally will help you break free of the negative opinions of others.

*The areas of impact that you are noted for are where your natural strengths lie,
regardless of what others think.*

Pivotal Moment in Action #5: We are all worthy of love, no matter what we've done or what has happened in the past. Take a moment to empower yourself and align to the gift of loving yourself and others. Like that moment where that pastor asked me to write my name in the place of *love*, find a quiet moment and write yours down with pen and paper.

(Your name) is patient.
(Your name) is kind.
(Your name) does not envy.

(Your name) does not boast.

(Your name) is not proud.

(Your name) does not dishonor others and is not self-seeking.

(Your name) keeps no record of wrongs and does not delight in evil but rejoices in the truth.

(Your name) always protects, always trusts, always hopes and always perseveres.

And the most important one:
(Your name) never fails.

*You are loved, and you are love.
Now is your moment to shine.*

Pivotal Moment in Action #6: We hold back emotions. We regret. We think about death as merely a possibility and not a guarantee. Think about who you would write a eulogy or a letter to if you heard they suddenly passed away? Who would you rush on a plane to see if they were on their deathbed? Like David, he gives love to people who are lonely, abandoned, and forgotten. Who could you give love to if you made the time? What would you want them to know?

*Your one act of love and kindness to one person might be
the one thing that saves them.
Be that reason, even if others disagree.*

Pivotal Moment in Action #7: Based on the widely shared story by Billy Graham, a tribal elder was teaching his grandson about life.

An old Cherokee grandfather is telling his grandson a story. "A fight is going on inside me," he said." It is a terrible fight between two wolves. One is evil—he is anger, envy, greed, arrogance, resentment, lies, and ego." He continued, "The other is good—he is joy, peace, love, hope, serenity, humility, kindness, empathy, generosity, truth, compassion, and faith. The wolves are fighting to the death.

Wide-eyed, the boy asks his grandfather which wolf will win. The old Cherokee simply replied, "The one you feed."

Which narrative do you play in your mind? Despite the difficulties you face in life, which wolf do you feed?

***The wolf you feed is the one who survives
long after you are gone. Let this be your legacy.***

Pivotal Moment in Action #8: Everyday miracles surround us. The universe is talking to us through people, songs, events and everything around you. Are you listening?

***Ask for clarity in your purpose, and the
answers will appear.***

Pivotal Moment in Action #9: It is said that if a memory makes you feel a negative or visceral response, it is unhealed and may be holding you back from your true potential. These are uncomfortable, unforgiven, or painful memories that replay constantly in your mind. No matter how terrible or painful it is, try writing it down. That was one of the hardest things I've ever had to do but in acknowledging the past, has helped me to release it. Let it hurt. Let it heal. Then, let it go.

Your truth might not be beautiful, but the beauty isn't in the pain—it's in rising above. Those who can sing after the rain don't forget the storm, but they know that the rainbow will eventually come.

Pivotal Moment in Action #10: Are there people in your life who hold you back, who constantly fuel fear in your dreams or decisions, or tell you that you're not ready? Rewrite how the story ends because no one's voice should be louder in your head than your own. If there is something you believe in, trust in your process, and take the steps necessary to make it happen.

*Trust in your process. From this moment on, no one's opinion or voice should be louder in your head than your *belief* in what you are capable of achieving.*

DAVID YOUNG'S JOURNAL OF TESTIMONIALS

"Let us build a house where love can dwell, and all can safely live—a place where saints and children tell how hearts learn to forgive." —Julia Fracassa

"We are blessed—in order that we might be blessings." —Bruce Boak, NY

"You are a shelter in more ways than one. I believe you guys truly are capstones." —Freddy Flogo

"From the very beginning, you opened up your world to me in a time in my life when I needed it the most. The journey that I am on is in its very beginning stages, and your example by living your own journey has given me great hope for the future. You have made your own life beautiful, and it shines very bright in this world." —Jesse Taylor

"By following your heart, connecting with your soul—your authentic self—and practicing being the best David you can be, you have connected so many people to each other and to themselves. In the world of disconnection, such connections are desperately needed to heal many wounds. I am so grateful that I learned how to face my fears and set them aside as I follow love. Every morning I can begin a new day knowing this is the day the Lord has made, let us rejoice and be glad in it. Isn't our new employer wonderful?" —Susan Schaffer

"Not a single one of us could even dream of being as compassionate, caring, and giving as you are." —Andrew (from Wilmington)

"What you do for this community and many, many strangers who just need shelter and an open heart is very refreshing and a great reminder of how just one individual can make such a huge impact." —Laura McSweeney

"We can't live alone. One has to help each other." —Hiroshi Okamoto, Japan

"I'm not sure what words to use to describe your home. It's a gem. A sanctuary, a reservoir, a garden growing people and dreams." —Heather Sedlacek

"Thank you for your generosity, your patience, your commitment to community, acts of loving kindness, for introducing me to the magical joy of bees." —Kaile

"David, your quiet example of generosity, selflessness, acceptance, hospitality, and faith is inspiring beyond measure. The world needs more David Youngs." —Dillon

"David, how do you begin to thank someone for giving you a home? Before I came here, I was camping out in a city park. And then your door opened to me, and I was greeted with a warm bed, a warm shower, and warm heater along with a new home. Thank you so much, David, and I wish you all the best in your efforts to continue to restore the Ninth Ward and provide people food or hope . . . What you're doing is so much more than just growing plants; you're growing a community. 'Start by doing what is necessary, then what is possible, and soon you will be doing the impossible.' —St. Francis of Assisi" —Steven

"Thank you for showing me that being the change you want to see is possible. I love everything that you are doing here and want to plant seeds and house bees (just like you) wherever I may call home." —Demi

"To provide a safe space in life's wayward path to our burgeoning generation while serving as gardeners in body and spirit for this place is a monumental righteousness. Feeding people's soul food for thought in the aftermath and cleanup of tragedy, integrating so many different causes and fostering the people's closeness to nature brings this place to the forefront against the numbness & isolation of our commercial society. This place of growth is a blessing to every life it touches. David, may your efforts be as fruitful and rich as the love and intention you have worked into the soil." —Duncan

"There is no doubt that the generative energy, love, and positivity that you regularly create in the world sustains a much-needed vanguard against a paradigm of scarcity that would have many believe it is not possible to open their lives up so much. Your actions create a balanced, positive system of reciprocity from which many could learn, and many have; it is quite clear." —Mars

"Thank you, David, for sharing the stories of how you've been helping to rebuild the community and telling me about your hopes for young people in Louisiana." —Stefanie

"You are a shining example of what it means to be a beacon of light in the world, to lead/serve by example, and to have your priorities anchored and guided by love." —Yeehni, Singapore

"Thank you for taking me in. Thank you for allowing me to work in the gardens. Thank you for breaking my heart with stories of some of the pain in the world and also for showing me that we are not helpless against it." —Rachel Cassar

"This has been by far one of the best experiences of my life! Thank you for inviting me in and showing me a different way of living." —Allison Brice, Tucson AZ/Washington, D.C.

"It has truly been a blessing . . . I have learned such diverse and valuable skills, I have met so many amazing people, and I have gotten to work for one of the most giving, compassionate, devoted, knowledgeable, and wise person I know. You have done great things with Capstone, or maybe I should say you have really allowed God to do great things through you and your work. I know you get a lot from the community, but I can clearly see how much better off the Lower Ninth Ward is because of you. Thank you for being my mentor, teacher, and friend. I am extremely thankful that I got to serve on this project. Not many get to see you support the youth in jail, but I wanted to make sure you know how valuable and powerful that ministry is. The boys love you and look up to you, and you do such a great job at mentoring them. I can't wait to see how Capstone touches more and more lives. I know Psalm 118:22 is often referenced but out of context. I really see how verse 23 fits all that you do here as well too. 'This is the Lord's doing; it is marvellous in our eyes.' Keep up the marvellous work." —Tyler

"It is absolutely clear to me that you have a servant's heart. You have little to give—you give freely with all that you have." —Kailah Netherchift, Florida

"Thank you for reminding me that every night and every day, we are reborn." —Nakylo

"We enjoyed working at the farm in the sun . . . we particularly enjoyed learning about the honey-making process and the worthwhile community-focused work that you do here . . . it is inspiring to see projects like yours in a world increasingly obsessed with consumerist individualism." —Kat + Jack, UK

"You have given me an unparalleled look into what New Orleans can be and what you have redefined for me what it means to lead a purposeful life." —Kate

CAPSTONE 118 FAQS

David has compiled a list of frequently asked questions. If you don't find your question answered here or would like further explanation, feel free to reach out to him directly at info@capstone118.org.

Can people send resources and books to the jail?
According to jail policy, all books must be sent directly from a publisher with the package listing the publishers stamp on the return. It must also be a soft paperback.
Resources can be sent directly to the Capstone address:

Capstone Shipping Address

P.O. Box 195

Arabi, Louisiana

70032-0195

USA

How can I donate? Are you a 501(c)(3)?

There are donation buttons on our website for general donations or our Adopt-A-Hive certificate program: Adopt-A-Hive or copy and paste: http://www.capstone118.org/Adopt-A-Hive.html You may mail a donation (checks only; please don't send cash in the mail) to:

Capstone

P.O. Box 195

Arabi, Louisiana

70032-0195

USA

If you have an in-kind donation or a special package to send, please contact us at capstone@capstone118.org to make proper arrangements.

Capstone is a 501(c)(3) non-profit corporation. You will receive an appropriate charitable contribution receipt for your donation.

Do you have paid positions?

Capstone is a volunteer-based organization, including myself (David). Currently there are no paid positions.

How can I volunteer?

We utilize many manners of volunteers to assist us in our work. We are working to refine how we handle volunteer requests and looking to create more volunteer opportunities available.

Can I visit or take a tour of Capstone?

We do allow visits or tours of Capstone. Some of our sites are available to see on your own. If you want to meet with us or see the aquaculture system, you will need to schedule a day and time. We don't have an office, and we often work on a different site each day.

Can I stay in my RV / camper or a tent while I visit or volunteer?

We do not have any type of camping accommodations on any of our properties. The city has been restrictive of people living in RVs or campers outside of a park since the last FEMA trailer was removed. The only camping option I know to suggest would be at Jean Lafitte State Park, about thirty minutes away, depending on traffic.

Is your food organic?

I have found out people have different understanding or perceptions of what the terms "organic" and even "Certified Organic" mean. We are not "Certified Organic." We grow our food as naturally as we can.

Do you use pesticides or herbicides?

As a beekeeper as well as a fish farmer, there are so many things—even those that are acceptable for "Certified Organic" use—that will harm the bees and fish. The only two naturally occurring pesticides I have ever used are 100% Neem Oil (properly diluted) and BT. With proper and considerate application, these will not harm the bees or fish in our aquaculture system.

However, I would like to point out that I am still using my original eight-ounce bottle I bought two years ago. This past year I have had good success using multiple applications of ladybugs for aphids. Hand-picking caterpillars has also proven very effective, and the fish love them as a snack.

I have not used any herbicides. I have considered using vinegar but haven't done so yet. Our most prevalent weed is Johnson Grass. One year, the roots had grown through all our potatoes. On the other hand, it's also our best source of feed for the goats. It is almost impossible to get rid of, so it becomes very labor intensive to control.

Can you give me gardening advice?

While I can answer general gardening questions, I am not able to give specific advice on your garden. There are many resources available. Most states have a land grant college or university with an ag extension office. These are great sources of information, and many have materials available online such as seasonal planting guides and nutrient guidelines for your specific area.

Growing in southern Louisiana is much different than growing in other areas of the US.

Do you grow year-round?

We are able to grow certain varieties of produce year-round. Most people who ask this are thinking about not growing in the winter. We grow lots of great produce all winter long. Our slow time is in the heat of July and August. Once it stays above 80 degrees at night, tomatoes won't even set fruit until it cools down again.

How did you get the property?

We have obtained the use of properties in many manners. I purchased the first lot at a very low price on a bank foreclosure. The lots for our largest orchard were bought by Capstone from New Orleans Redevelopment Authority after winning a pitch competition they co-sponsored. We have leases that vary from five or ten years to indefinite. These property leases come from other organizations on lots they are not ready to rebuild on or are too small to rebuild on, private owners who are not ready to rebuild yet, and one out of country owner who bought it as an investment and fell in love with what we do. We also have some that we take care of because it adjoins another lot we use.

Why do you have the animals? Are the animals raised for meat?

Currently, we have chickens, goats, and ducks. I can't say that none of our animals have ever been butchered, but they are not raised for that purpose. Our animals are an important source of fertilizer for the gardens. Goat manure is a cold manure and can be put directly on plants without burning them. The chicken and duck manure are hot manures and need to be composted along with the Johnson Grass we use for bedding and feed. The goats also do a good job of lawn mowing some of our areas. I put a temporary fence around the lot beside our house and let the goats graze there a few times a week. We haven't mowed that lot this year. The animals also provide a connection to the community to allow them an opportunity to interact with animals they may not otherwise see.

Why do you have bees?

I became a beekeeper out of necessity. When you think about the loss from Katrina, many people don't think of the loss of wildlife under fourteen feet of water for three weeks. One first responder said the absence of birdsongs or other sounds in the morning was eerie. Honeybees are already struggling for their survival with losses around 40% each year.

My first two years of gardening, I counted a total of six honeybees in the garden. I wasn't getting the pollination I needed to grow produce. After many months of research and trying to find someone else to put a hive on my garden, I was presented my first hive of bees. When Hurricane Isaac came through, I loaded up my beehive in my Explorer, and they left with me. I continued to expand the number of hives I kept, and after a few years, found there was a demand for good local raw honey that could provide a source of revenue to support our mission.

In 2015, we made a large investment to enter retail market sales. This allows us to sell in grocery stores and restaurants. Capstone Raw Honey has been received well in that market. We have continued to expand and now have sixty hives in the Lower Ninth Ward.

Do you ship Capstone Raw Honey?
No, unfortunately, we do not.

What is the 118 in Capstone118?
Psalms 118:22 says, "The stone the builders rejected has become the capstone."

Capstone is not a faith-based organization or a church. However, it is an organization that I founded on the understanding of living out my faith as I believe I should. I found this segment of commentary online at **Think Hebrew**—Christian Reflections on a Jewish Messiah *that talks about building the Temple:*
The very first stone to be delivered was the capstone, but that's the last stone needed in construction. So the builders said, 'What is this? This doesn't look like any of the first stones we need. Put it over there for now.' *Well, years went by and the grass grew over the capstone and everyone generally forgot about it. Finally, the construction was done and the builders said,* 'send us the capstone' and the word came back from the quarry 'we already did'. They were confused. Then someone remembered what they had done with the very first stone sent to them. It was taken from its lowly position among the overgrown weeds where it had been forgotten, and it

was honored in the final ceremony to complete the temple. Thus the scripture says, 'The stone the builders rejected has become the capstone.'

Myself and others in consultation felt this was very applicable to our mission and specifically to those we serve in the Lower Ninth Ward and the larger community.

Many ask, "How can I start doing something?"

When I started doing volunteer work and traveled more, I saw opportunities to help in almost every place I stopped in. For me, it was a matter of walking beyond the common areas, slowing down enough to actually become part of what was around me, rather than simply pass through. Listen to what people say their needs are and assist within your means.

My garden started as four feet by fourteen feet, and if people wanted to help, all they had to bring were the seeds or plants they wanted to put in the ground. When I spent $100 during the first two years, that was all I had to spend. Even as we have expanded, we operate with very limited funding. Helping others became easier for me once I realized I didn't need to change the whole world. I could make a difference in my part of it, then let the rest of the world see

RESOURCE GUIDE

The First Step Act Resource Guide
If you or a loved one have recently returned from federal prison, this guide is created for you.
Dream Corps' #cut 50 program, a bipartisan effort to cut crime and incarceration across all fifty states, has partnered with Root & Rebound to produce a 165-page long re-entry guide and resource directory designed specifically for community members returning home under the First Step Act. This guide will help you or a loved one obtain official identification, find work, know your fair chance employment rights, and so much more.
https://www.thedreamcorps.org/
The First Step to Second Chances Guide can be found at the link below:
https://d3n8a8pro7vhmx.cloudfront.net/rebuildthedream/pages/23318/attachments/original/1563549623/First_Step_to_Second_Chances_guide.pdf?1563549623

Homeboy Industries
Homeboy Industries is the largest and most successful gang intervention, rehab, and re-entry program in the world and has become a model for other organizations and cities. Founded by Father Greg Boyle in 1992, the program assists

high-risk youth, former gang members and the recently incarcerated with a variety of free programs, such as mental health counseling, legal services, tattoo removal, curriculum and education classes, work-readiness training, and employment services.

What began as a way of improving the lives of former gang members in East Los Angeles has today become a blueprint for over 250 organizations and social enterprises around the world, from Alabama and Idaho, to Guatemala and Scotland. The Global Homeboy Network is a group of like-minded organizations committed to impacting the lives of those in their communities. https://homeboyindustries.org/

Books for Kids Foundation with Plenty.org
Capstone is proudly partnered with Books for Kids Foundation, who provides donated books to children and teens in the Lower Ninth Ward, through the support of Plenty.org. Plenty is a not-for-profit organization that was created to help protect and share the world's abundance and knowledge for the benefit of all. Plenty delivers books to Capstone for Pre-Kindergarten, Kindergarten to Grade 8 Academy and High school for all the local community kids to take. If you would like to donate books to the youth in the Lower Ninth Ward or to the teens in jail, please ship directly to:
Capstone
P.O. Box 195
Arabi, Louisiana
70032-0195
USA

Reboot and Rebound

The Re-entry Planning Toolkits are designed for different audiences, with the goal of helping people with arrest and conviction histories and those who support them to develop re-entry plans collaboratively. The toolkits give access to practical tools and know-your-rights information so that systems-impacted people can have the chance to thrive, not just survive, after incarceration.

https://www.rootandrebound.org/resources/reentry-planning-toolkit/

ACKNOWLEDGEMENTS

The dream of writing this book became a reality with the help and support of many special people.

First, I want to thank David Young for trusting me with your incredible story. One of the most important lessons you've taught me is that when we walk in faith with God as your guide, *anything and everything* is truly possible. You live every day with inspiring conviction, and that's what changes the world.

Thank you to Kary Oberbrunner and the entire Author Academy Elite team. You have shown me what 'showing up filled up' means, and have been the strength I needed when writers block took over. You inspired me to keep going, to trust in the vision and become the author we are all meant to be.

I couldn't have done this without my editors Jane VanVooren Rogers and Tina Morlock. Your wisdom, insight and guidance throughout each step helped to bring this story to life. Thank you for believing in this project, and in me.

Thank you to Debbie O'Byrne, JetLaunch, Daniel Anupol, Mike & Linda Trabulsi, Abi Shaktivel for your creative support. We were all meant to cross paths for a reason, and I'm beyond grateful for that.

To the inspiring William Divine, Jeffrey Neubauer, Kevin E. France, for your belief in this mission and message. The world needs more leaders like you.

Thank you to my mentor Dr. Len Karakowsky. Through your guidance, you've shown me what a great teacher looks like. I am lucky to have been your student.

To my loving extended family and friends here in Toronto, California and around the world, you are the village I grew up with since I was young and I carry you with me in my heart.

To my brothers, RJ Bautista and Ross Jr. Bautista, thank you for walking with me as my true day ones. I appreciate all the talks and patiently offering your feedback and advice. I learn from you, I lean on you and I appreciate you always.

Thank you to my mom and dad who never let me feel the weight of chasing my dreams. You gave me love, and you gave me wings. And with those two things, you've given me everything. I can fly because of you.

To my husband Neil, for your endless support in everything I do, I love and thank you. You have given me the greatest gifts with our boys, Christian and Connor, and this journey of a lifetime together.

Lastly, I'm thankful for every answered prayer along the way and can honestly say that when you let God into your life, miracles can happen. Thank you for walking this journey with me, let's continue to change the world together.

With love,
Kristle

Connect With Me

I'd love to hear your thoughts and experiences
with your Pivotal Moments
Please connect with me at www.kristlebautista.com

Global change *is* possible, starting with you.

ABOUT THE AUTHOR

 Kristle Bautista, *MHRM, CHRL,* is a Course Director at York University with the Faculty of Liberal Arts and Professional Studies. Recognized for excellence in teaching, she is a natural story-teller and lead panelist for Human Resources. Her global HR advisory, mentorship and training experience spans across the Aerospace, Business Consulting, Not-for-Profit, Food and Consumer Packaged Goods industries. She is a natural introvert and a true champion for the quiet ones too.

Kristle lives in Toronto, Canada with her husband Neil, and their two boys Christian and Connor.

ENDNOTES

1   www.capstone118.org

2   https://www.facebook.com/RobGreenfield/videos/93365
    0420095370/?vh=e

3   https://www.livescience.com/22522-hurricane-katrina-
    facts.html

4   https://www.worldvision.org/disaster-relief-news-stories/
    2005-hurricane-katrina-facts

5   https://www.nationalgeographic.com/environment/
    natural-disasters/reference/hurricane-katrina/

6   https://www.areavibes.com/new+orleans-la/most-
    dangerous-neighborhoods/

7   Mumia Abu-Jamal Title: Have Black Lives Ever Mattered?
    Publisher: City Lights Publishers Publication Date: 07/04/17
    ISBN: 9780872867383 Trade Paperback (English)

8   https://www.theadvocate.com/baton_rouge/news/article_
    4dcdfe1c-213a-11ea-8314-933ce786be2c.html

9   https://en.wikipedia.org/wiki/Jeffrey_A._Neubauer

10 https://www.justice.gov/usao-edla/pr/c-ray-nagin-former-
    new-orleans-mayor-convicted-federal-bribery-honest-
    services-wire

11 https://www.nytimes.com/2014/02/13/us/nagin-corruption-verdict.html

12 https://en.wikiquote.org/wiki/Field_of_Dreams

13 https://quotefancy.com/quote/843318/John-Green-The-darkest-nights-produce-the-brightest-stars

14 https://www.telegraph.co.uk/science/2017/08/06/loneliness-deadlier-obesity-study-suggests/

15 http://www.oprah.com/own-super-soul-sunday/shaka-senghors-life-changing-epiphany

16 https://en.wikipedia.org/wiki/Shaka_Senghor

17 https://time.com/5405158/the-true-history-of-americas-private-prison-industry/

18 https://www.nydailynews.com/entertainment/movies/danny-trejo-celebrates-48-years-sober-article-1.2765856

19 https://en.wikipedia.org/wiki/Two_Wolves

20 https://variety.com/2015/film/news/oil-industry-exec-launches-film-production-company-1201567223/

21 https://www.imsdb.com/scripts/Good-Will-Hunting.html

22 https://newyorkhistoryblog.org/2017/04/the-mysterious-death-of-the-angel-of-sing-sing/

23 https://www.psychologytoday.com/us/blog/workings-well-being/201708/heal-trauma-work-the-body

24 https://en.wikipedia.org/wiki/First_Step_Act

25 http://www.capecharlesmirror.com/news/first-woman-released-under-the-first-step-act/

26 https://www.candoclemency.com/catherine-toney-20-years/

27 https://www.artsy.net/article/artsy-editorial-painting-helped-wrongfully-convicted-man-prison

28 https://www.nytimes.com/2019/08/27/arts/music/meek-mill-free.html

29 https://www.theguardian.com/music/2019/jul/24/meek-mill-conviction-overturned

30 https://poets.org/poem/if

31 https://en.wikiquote.org/wiki/Field_of_Dreams

Printed in Great Britain
by Amazon

45009977R00097